Leech

Animal

Series editor: Jonathan Burt

Already published

Ant Charlotte Sleigh · *Ape* John Sorenson · *Bear* Robert E. Bieder
Bee Claire Preston · *Camel* Robert Irwin · *Cat* Katharine M. Rogers
Chicken Annie Potts · *Cockroach* Marion Copeland · *Cow* Hannah Velten
Crow Boria Sax · *Dog* Susan McHugh · *Donkey* Jill Bough
Duck Victoria de Rijke · *Eel* Richard Schweid · *Elephant* Daniel Wylie
Falcon Helen Macdonald · *Fly* Steven Connor · *Fox* Martin Wallen
Frog Charlotte Sleigh · *Giraffe* Mark Williams · *Hare* Simon Carnell
Horse Elaine Walker · *Hyena* Mikita Brottman · *Kangaroo* John Simons
Leech Robert G. W. Kirk and Neil Pemberton · *Lion* Deirdre Jackson
Lobster Richard J. King · *Moose* Kevin Jackson · *Mosquito* Richard Jones
Otter Daniel Allen · *Owl* Desmond Morris · *Oyster* Rebecca Stott
Parrot Paul Carter · *Peacock* Christine E. Jackson · *Penguin* Stephen Martin
Pig Brett Mizelle · *Pigeon* Barbara Allen · *Rat* Jonathan Burt
Rhinoceros Kelly Enright · *Salmon* Peter Coates · *Shark* Dean Crawford
Snail Peter Williams · *Snake* Drake Stutesman · *Sparrow* Kim Todd
Spider Katja and Sergiusz Michalski · *Swan* Peter Young · *Tiger* Susie Green
Tortoise Peter Young · *Trout* James Owen · *Vulture* Thom Van Dooren
Whale Joe Roman · *Wolf* Garry Marvin

Leech

Robert G. W. Kirk and Neil Pemberton

REAKTION BOOKS

For leeches, everywhere

Published by
REAKTION BOOKS LTD
33 Great Sutton Street
London EC1V 0DX, UK
www.reaktionbooks.co.uk

First published 2013

Printed and bound in China by C&C Offset Printing Co., Ltd

British Library Cataloguing in Publication Data
Kirk, Robert G. W.
Leech. – (Animal)
1. Leeches. 2. Leeches – Therapeutic use.
I. Title II. Series III. Pemberton, Neil.
592.6'6-DC23

ISBN 978 1 78023 033 7

Contents

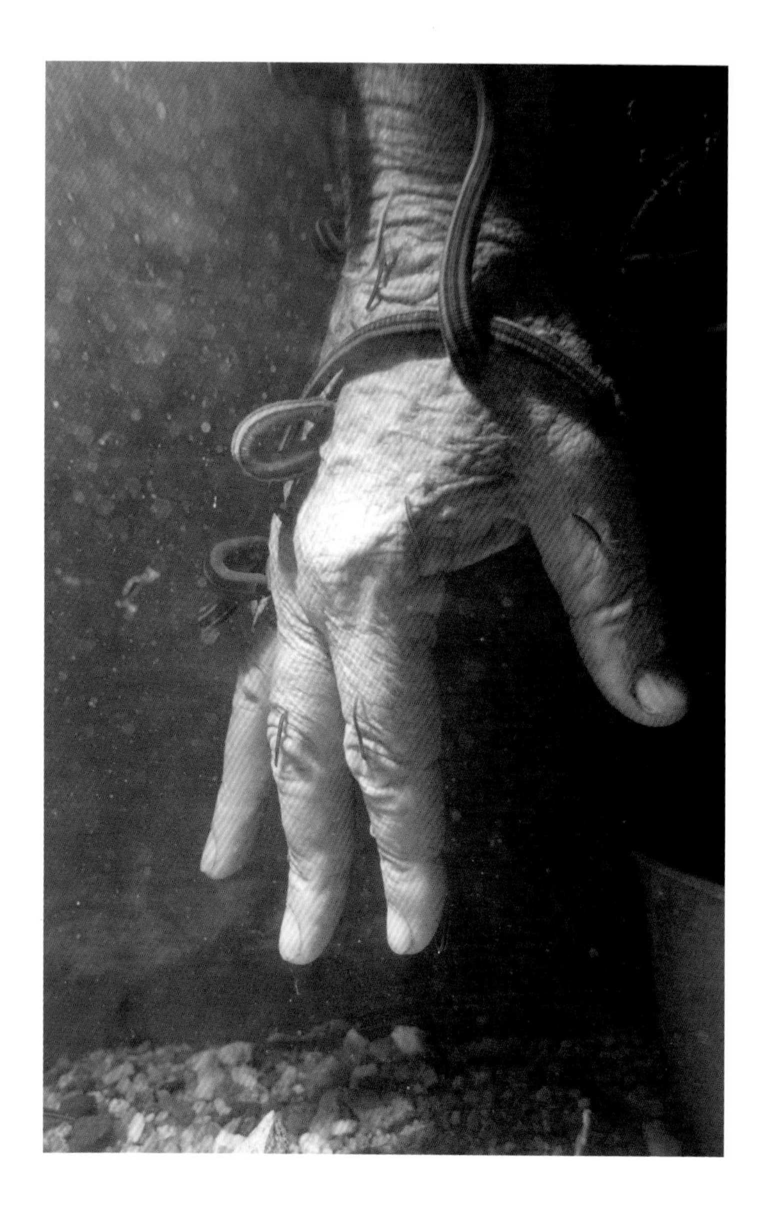

Introducing Leech

'For the sake of the leech I lay here by this swamp like a fisherman, and my outstretched arm had already been bitten ten times when a still fairer leech came after my blood: . . . the great conscience-leech, Zarathustra!' . . .

'Who are you?' he asked . . .

'I am the *conscientious in spirit,*' answered the one who had been asked . . . 'Rather know nothing than half-know many things! Better be a fool on one's own account than a wise man in the opinion of others! I get to the ground of things – what does it matter whether it be great or small? Whether it be called swamp or Heaven? A handsbreadth of ground suffices for me . . . In the right science of conscience there is nothing great and nothing small.'

'Then perhaps you are the connoisseur of the leech?' asked Zarathustra. 'And you pursue the leech down to the ultimate grounds, you conscientious one?'

'O Zarathustra,' answered the one stepped on, 'that would be something immense: how could I understand such a thing! But what I am master and connoisseur of, that is the leech's brain: that is my world!'

Friedrich Wilhelm Nietzsche, *Thus Spoke Zarathustra* (1883–5)

In this book we want to achieve something immense, to introduce you to the leech. By its end we hope you will have become a connoisseur not just of the biology and brain of the leech but of its world. A world, we argue, that has always already been our own. A horror and a healer, the leech is one of nature's most tenacious yet mysterious animals. Possessing an unassailable hold on the human body and on culture, few animals feature so unexpectedly yet consistently in human history. Armed with razor-sharp teeth, a muscular body and possessing a piercing bite, leeches are capable of drinking many times their weight in blood. One might think this would make them a most unlikely animal to turn to in times of ill health; yet this is precisely what human cultures all over the world have been doing for millennia. Indeed, leeches may justifiably claim to be one of humanity's oldest and most enduring, albeit peculiar, companions.

In many ways leeches can be a very difficult animal to get to know. Leeches are worms, and as such are as different to us as an animal can be. Distinguishing the head of a leech from its tail can be a challenge, and even when the head is found it is not easy to recognize a face as conventionally imagined. Although preferring aquatic environments, leeches can be found in a vast diversity of habitats. Though most are exclusively blood-feeders, some species of leech consume the flesh of small animals, particularly invertebrates. Humans have lived intimately with leeches throughout recorded history, yet the private life of this creature has remained mysterious. For every occasion in which we have lived, worked and cared for leeches, there have been parallel moments when they have inspired disgust, fear and repugnance. In the entangled history of human and leech, we have repeatedly accepted and rejected, partnered and separated, loved and loathed each other, each generation thinking the relationship to be new and forgetting that our predecessors have been through it all before. What a human story. Or is it a leech story? Perhaps we should suspend, for a moment, a strong distinction between the two.

There are estimated to be over 650 leech species and still more to be discovered. But what is a leech? In general terms, a worm. More specifically, a segmented worm or 'annelid'. Leeches are identified by their possession of suckers, one at each end of their wormy bodies. Whilst the unique biology of leech species is interesting in itself, it becomes yet more fascinating when understood alongside the distinctive cultural roles leeches have played in human societies over time. Our biology shapes but does not determine what we are as humans. Nor can biology alone determine the identity of a leech. Have you ever been called a leech? Chances are you would not like it if you were. Today, to be called a leech is to be associated with a parasite, a style of living that consumes without giving. When the British parliamentarian Ken Purchase accused 'leeches' of killing British industry in 1998, it was 'non-producing' bankers whom he was calling to task. The rhetorical use of leech as parasite works to other, to define boundaries between 'us' and 'them'. In this book we will resist this line of thinking because the language of parasitism is already laden with negative values. Instead, we wish to blur boundaries, revealing, for example, how leeches are defined not just by biology but by human culture. In recent years leeches have been anthropomorphized primarily to embody the worst of humanity, our darkest, most aggressive and avaricious tendencies, the aspects of our characters that we like to think are *not human*.

However, we would suggest that giving the name 'leech' to such horrific human creations as bankers, vampires and other monstrosities is unfair to the wormy creatures that also go by that name. In other times and other places, leeches, even when humanized, have not been viewed as monstrous. On the contrary, for many centuries leech was an alternative term for a physician. In Old English, 'leech' (derived from the Germanic 'laece') meant 'to heal', whereas the word for the creature derived from the Middle Dutch, also 'laece', meaning 'worm'. Old English medical texts were known as 'leech books', such as those collected by Thomas Oswald Cockayne as *Leechdoms, Wortcunning and Starcraft of Early England* (1864). Yet, as we shall see, leeches were literally healers. Despite the two words having separate etymological origins, their meanings overlapped. For many centuries, leeches (the animals) were used by leeches (the physicians) to leech (heal). Thus leeches were used to leech by leeches!

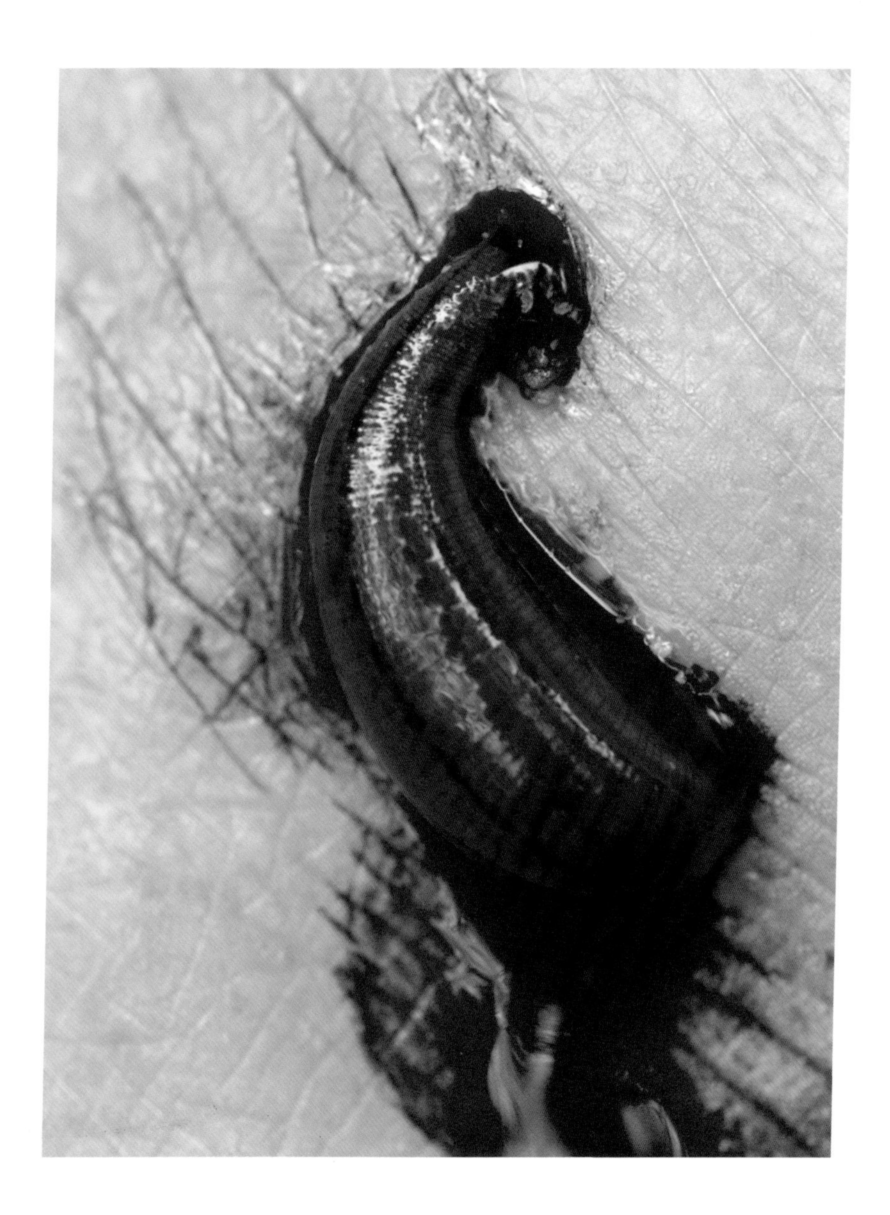

Woman applying leeches to her forearm from a jar, from Guillaume van den Bossche, *Historia medica* (1638).

How do we make sense of these ambiguities? How can the leech be both healer and horror? On 24 March 2008, appearing on the *Late Show with David Letterman*, Demi Moore described how she maintained her youthful age-defying appearance:

> I feel like I've always been someone looking for . . . things that optimize your health and healing . . . I was in Austria doing a cleanse and part of the treatment was leech therapy. These aren't just swamp leeches though – we are talking about highly trained medical leeches.

Moore believes that leeches detoxify her blood, optimize her health and, hinting at the intimacy she shared with these animals, went on to explain how 'leeches don't like hair, they much prefer a Brazilian'. Here, centuries-old associations linking leeches to bodily youth, longevity, cleansing, healing, beauty and sexuality

Bloodsucking leech!

are revived and woven anew into twenty-first-century celebrity culture. Moore also stated that the body and spirit are not so easily separated: it was not just that her blood was detoxified, she *felt* detoxified, demonstrating that body and belief, corporeality and culture, are always already interwoven. Leeches, therefore, cleanse the human condition, not just in a bodily sense but also in a cultural one. Greed, aggression, insatiable hunger, avarice, uncaring self-centredness, murderous bloodthirsty intent: all the darkest and worst of human nature can be poured through metaphor into the leech. These all too human characteristics, which *to be* human we must *deny as being* human, are thereby disposed. The leech purifies the human condition by becoming, in our imaginations at least, all that we do not want to be. Quarantining our intolerable Hyde in the body of the leech allows us to believe that to be human is to be Jekyll alone. Just as the leech draws out and contains bodily impurities, so too have these animals acted as cultural containers for all the bad in our character and soul. This can help explain how and why leeches are both loved and loathed.

But the cleverest trick is that this relationship can only work if we *believe* that the leech is the parasite. Leeches must be parasites to hide what it is we are. Leeches must be the very symbol of their oppressor. The truth is quite the opposite: it is humans that are the parasites. In this book we resist understanding the leech through the category of parasite. Instead, we explore how leeches might be understood to play a symbiotic role in human culture. We show how humans and leeches have lived together, supported one another, and shared mutually beneficial relationships. We examine how, through our interactions over centuries, humans and leeches have made and remade each other countless times. The leech is revealed to be one of our oldest non-human companions, a partner in the making of our past, present and future.

Various examples of freshwater leeches, including the Glossiphoiidae, carrying their young on their backs.

The nature of leeches comes from their place in culture, their place in history; it is neither determined by their biology nor essential to their being. Nature is always embedded in culture and history. We weave meaning from already existing tangled webs. When you were a child it is likely you were a schoolchild. But you were not born a schoolchild. That identity grew from your place within a wider cultural environment. You are no longer a schoolchild, or if you are, one day you will not be. Yet the echo of that schoolchild will shape you for life. So it is with leeches. Twenty-first-century leeches are no longer what they were in the nineteenth century, or the twelfth. But the echoes of those older leeches, the roles they played in culture, the work they performed in society, reverberate still in our collective memory. In the pages that follow, we hope to awaken that memory. We will introduce you

Group of *Hirudo medicinalis* (medicinal leech).

to leeches that heal, leeches that hurt. Leeches that will suck your blood and happily take up home in your nose. We will encounter leeches that live in pools, rivulets and streams, leeches that prefer to crawl on land, leeches that live in the sea, leeches that climb trees and even leeches that fly (or at least are specialists in falling). We will find leeches that work in laboratories, in hospitals, in literature and art. We will meet leeches you may never wish to meet again. We will learn of leeches that can predict the weather. Leeches that have become machines. Leeches that have taken on human form, and leeches that have become monsters. But most of all, we want to introduce you to the leech who you might think of as a companion. Perhaps, we hope, a friend.

1 Natural Leech

Is it some primordial memory that causes us to respond to leeches with alarm and disgust? No. On the contrary, for much of our history leeches have occupied a valued position in human culture and enjoyed a far more intimate relationship with humanity than their less intimidating wormy cousins. If asked to imagine a worm, the chances are you would picture in your mind an earthworm. Though it may not be immediately obvious, earthworms perform important work. Charles Darwin was the first to describe how earthworms maintain the fertility of soil. Observing their habit of 'crawling over each other's bodies', he speculated on their emotional capacity, suggesting earthworms could possess 'a trace of social feeling'.[1] We should be thankful for earthworms, Darwin concluded, quietly getting on with their work deep beneath our feet. But not all worms appear quite so blameless. Leeches, for example, are also worms, yet in Western culture today they rarely receive thanks for their labour. As anyone who has encountered these distant evolutionary cousins of the earthworm will know, leeches will gladly set to work less below your feet than upon them.

WHAT MAKES A LEECH A WORM?

The eighteenth-century Swedish botanist, zoologist and sometime physician Carl Linnaeus was the first to place the leech within

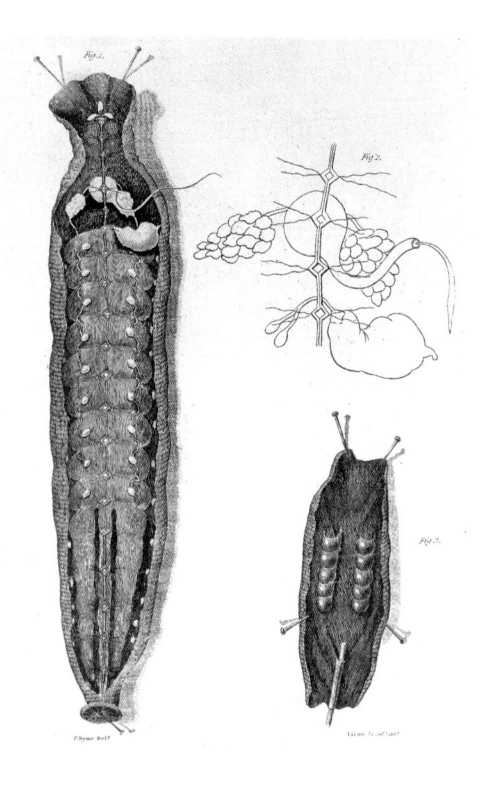

A dissected medicinal leech, from James Rawlins Johnson, *A Treatise on the Medicinal Leech* (1816).

a huge heterogeneous class named Vermes ('worms'). Leeches belong to the taxonomic phylum Annelida (latin *annelida*, 'ringed ones'), created in 1818 by the French naturalist Jean-Baptiste Lamarck for all worms whose bodies were made up of a number of small rings.[2] This collection of wormy creatures can be recognized by their elongated bodies made up from ring-like segments, separated from each other by shallow constrictions clearly visible on the external body. All annelids possess bodies made up of three areas: the prostomium, or head, containing brain, mouth and sense organs; the long, ringed central trunk; and the pygidium,

Ancient leeches, from Johannes Jakob Scheuchzer's *Physica Sacra* (1731–3), which provided a scientific account of biblical events.

containing the important though often elusive anus. This instantly visible 'annulated' effect provides a simple way to distinguish annelids from their equally wormy kin.

Unlike most mammals, worms lack an apparent face, and their bodies have no discernible up and down, nor an instantly obvious front and back. It is tempting, therefore, to overemphasize their difference. In the nineteenth century, for instance, it was thought that leeches did not have an anus but rather excreted waste via the skin through a process called 'transpiration'.[3] The leech anus is particularly difficult to locate not only due to its minute size but also its unusual positioning: where the anus would be in other worms, the leech possesses a posterior sucker. Early students of leech anatomy, being unable to see an anus, were misled, mistaking the unusual thread-like substance that rubbed off the external parts of the leech body for excrement. This mucus-like substance is, in fact, more akin to human urine. Leeches expel fluids through excretory glands called metanephridia which, in a comparable manner to the human passing of urine, allows leeches to control the levels of vital substances (such as salt) within their bodies. The belief that leeches uniquely lacked an anus was related to their special position within early nineteenth-century medical culture. As we will learn in the following chapter, leeches were widely used to draw out and neutralize impurities from the bodies of others, thus it made sense that they should have a unique physiology. As the leech became better studied and tools of investigation improved towards the end of the century, it was realized that, like the majority of annelids, leeches possessed a relatively simple yet complete digestive tract. This takes the form of a tube passing through each segment (or 'septa') from mouth to anus, which, of course, is necessary for expelling solid waste.

One structural consequence of segmented bodies is that annelids have no need of hard skeletons. Instead they possess a 'hydrostatic' skeleton, consisting of a fluid-filled cavity enclosed by muscles. Muscle activity is used to change the pressure on the fluid, altering the creature's shape. This produces the characteristic wormy movements so well adapted for burrowing and swimming. Such flexibility, however, comes at a cost. Annelids rarely appear in the fossil record because without a hard skeleton there is nothing solid to be preserved, meaning we know little of their evolutionary history. Explaining how and why annelids developed their segmented bodies remains the 'holy grail' of twenty-first-century evolutionary morphology.[4]

Although Lamarck named the annelids, separating them from arthropods (such as spiders and centipedes) and associating them with molluscs and crustaceans, others disagreed. The French naturalist Georges Cuvier, a contemporary rival of Lamarck, believed annelids and arthropods to be closely related since both possessed segmented bodies.[5] For good measure, Cuvier tweaked Lamarck's definition of Annelida so as to mean worms possessing red blood (particularly resonant for the leech). Cuvier's disagreement with Lamarck was not motivated solely by the family relationships of worms. Nor was the colour of leech blood the primary issue. Lamarck and Cuvier differed on the most important question of their day: evolution. Where Lamarck was a committed evolutionist, Cuvier was vehemently opposed to the theory. A generation later, when the eminent German biologist Ernst Haeckel came to the question of worms and their kin, he followed Cuvier. Observing that annelids and arthropods shared similar physical characteristics in possessing segmented bodies, Haeckel concluded they must be related. He did so, however, within a post-Darwinian evolutionary framework. This difference is subtle but important.

After Darwin published his theory of evolution in 1859, nature could be ordered in a new way, by its evolutionary relationships or 'phylogeny'. Biologists created complex phylogenetic trees, much as a genealogist might produce a family tree. Despite this critical conceptual change, in practice, Haeckel's classification continued to rely on the comparison of morphological traits as a means to identify evolutionary kinship. Whilst the process of mapping relationships between species was little different, the meaning of those relationships had forever changed. To think in evolutionary terms was to posit a tangible historical kinship;

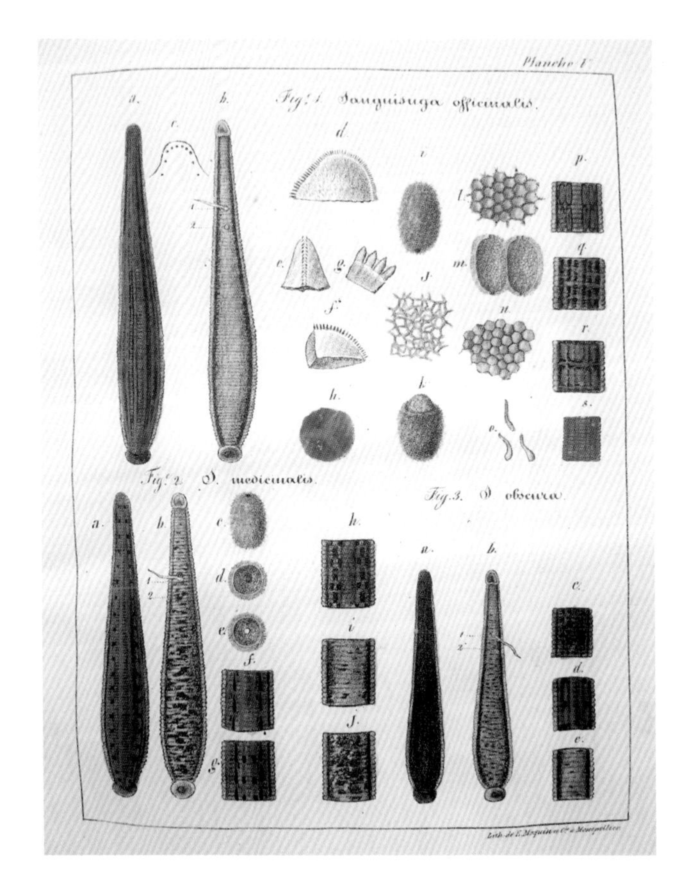

Leech anatomy, from Alfred Moquin-Tandon, *Monographie de la famille des hirudinées* (1846).

species were, for example, ordered with reference to a common ancestor. Consequently, classification could help explain how, why and when some of the more complex and curious physiological and ecological characteristics had come to be.

This is particularly important for annelids, and for none more so than leeches due to their absence from the fossil record. There are only two reported discoveries of fossilized leeches – of

Epitrachys rugosus (1869) and *Palaeohirudo eichstaettensis* (1970). Neither is a well-preserved fossil. They do not demonstrate the telltale caudal sucker used by the leech for attaching to surfaces, nor is it clear if they possessed a segmented body. As a consequence, there is little evidence as to when leeches diverged from ancestral worms. However, since both fossils date to the late Jurassic period (about 1.5 million years ago) it is generally accepted that the common ancestor of all leeches would have emerged around this time, alongside the earliest mammals. Without the ordering theory of evolutionary kinship, it would be very difficult to understand any more about the history of leeches. However, within evolution, morphological similarities between living creatures can be interpreted as indicators of kinship. The history of leeches can therefore be pieced together from their biological similarities to other creatures.

There are estimated to be 12,000 different species of annelids adapted to live in oceanic, freshwater and terrestrial environments. Within the traditional Linnaean classification system, Annelida were sorted into three groups. The most diverse, Polychaeta, is almost entirely made up from marine creatures, such as lugworms, ragworms, and fan worms, totalling some 8,000 species. Oligochaeta, including all earthworms alongside some marine creatures, has at least 3,500 members. Finally, the smallest group, reserved for leeches, is the Hirudinea, numbering an estimated 650 species. However, in recent years this long-established classification has been questioned. Since Darwin, a more objective method for establishing evolutionary kinship had been sought because morphological difference often lay in the eye of the beholder. Around the mid-twentieth century, for example, the technique of comparing obvious morphological traits was partly superseded by the study of embryological similarities. Today, molecular modes of rendering differences between species have become dominant. Using the latest technologies of DNA analysis the modern science of cladistics aims to order nature within 'clades' (consisting of a single ancestor and its entire descendant species). Though cladistics has led to resurgence in the production of phylogenetic trees (now sometimes called 'cladograms'), the results of DNA analysis have not been straightforward. In many cases, molecular findings have failed to correspond with older methods of classification, casting long-established taxonomic relations into doubt. On the basis of DNA similarities, leeches, the Hirudinea, are now thought to form a subclass of Oligochaeta that in turn might best be seen as a subgroup of the Polychaeta. At the same time, the Pogonophora (or 'beard worms'), Echiura ('spoon worms') and Sipuncula ('peanut worms'), all traditionally categorized elsewhere, are now also thought to be subgroups of the Polychaeta, thus becoming distant relations of leeches.

Confused? So is the field of invertebrate biology. The most recent textbooks acknowledge that the evolutionary kinship of worms is in flux. The strong distinction between the Oligochaeta (earthworms) and the Hirudinea (leeches) has now been abandoned. At least temporarily, their relatedness is now emphasized, with both placed within a class called the Clitellata. Presently, then, earthworms form the closest evolutionary relatives of leeches. Beyond that the jury remains firmly out. Recall the debate regarding whether annelids were related to molluscs as Lamarck believed, or arthropods as Cuvier held? Due to both possessing segmented bodies, Haeckel sided with Cuvier and so for over a century worms and arthropods were considered to be evolutionary kin. In the twenty-first century, however, Lamarck has been vindicated. DNA evidence suggests that molluscs are the most recent common ancestors of annelids.[6] As new ways of reading wormy kinships reshuffles evolutionary relationships yet again, the immediate family of leeches remains uncertain. But before we feel too bad for leeches we should remember that they are never lonely. Their family has countless members. Moreover, there is nothing leeches enjoy more than courting intimate relationships with other species.

WHAT MAKES A LEECH A LEECH

Mention a leech and two things leap to mind – blood and sucking. The French word for leech, *sangsue,* derives from the Latin *sanguisuga* conjoining *sanguis* ('blood') with *sugere* ('to suck'). Yet not all leeches feast on blood. What distinguishes leeches from other worms is their possession of anterior and posterior suckers, used as their primary means of locomotion. Despite being limbless, leeches are highly mobile thanks to their suckers. In many species the anterior sucker, surrounding the mouth and generally the smaller of the two, is precisely adapted for blood feeding. Powered by a unique bodily structure, the larger, posterior sucker plays the major role in locomotion. Unlike other worms, the leech, though retaining the appearance of segmentation, does not in fact possess the segmented body typical of annelids. The bodies of most worms consist of individual segments each possessing a pair of fluid-filled cavities (the 'coeloms') which when manipulated by muscles enable movement. Leeches differ in possessing only a single pair of coeloms running uninterrupted through the body. In addition to being the two main locomotory vessels, this unique physiological feature has evolved to function as a circulatory system, acquiring arteries, veins and capillaries. Leech bodies are also distinctive in that they have lost the hair-like bristles (chaetae), commonly found on annelids. These peculiarities are bound up with the functional changes that have given leeches their unique method of locomotion. Unlike worms, leeches do not propel themselves through peristalsis-style burrowing. Instead, leeches use their posterior sucker as a base from which a wave of muscular contractions begins, extending the body and propelling it forward. By attaching their anterior suckers to a surface leeches can, through further muscle activity, contract their bodies, dragging themselves forwards. These movements produce a caterpillar-like manner of travelling, making the leech a surprisingly adept land animal. Nonetheless, it is in aquatic environments that leeches come into their own. By flattening and manipulating their bodies into wavelike patterns, leeches are capable of swimming at speed and with an elegance few other creatures can rival.

The leech world is little understood. For instance, leeches are adept at orientation, but how they achieve this is somewhat of a mystery. Despite possessing apparently simple sensory systems, leeches are highly alert to their environment, able to seek out potential hosts from some distance. Early naturalists spent much

The horse leech (*Haemopis sanguisuga*) feeding on a slug.

ink speculating on their capacity to see, smell, taste, hear and feel. James Rawlins Johnson, a nineteenth-century physician, believed leeches possessed the full complement of senses. He assured his readers that counter to appearances, microscopic investigation revealed that leeches possessed ten deep black eyes. Leeches' sense of touch, Johnson continued, was mostly situated in the lips and their sense of taste located in the oesophagus. His reasoning was derived less from anatomical research than his belief in the continuity of all living things. On the sense of hearing, he wrote, 'the closest anatomical inspection does not advance us a single step in determining the situation of the organ' yet 'there can be no doubt that it is granted to the Leech in common with the numerous tribe of animated beings'.[7] The sense of smell was similarly assumed to exist. Johnson was determined to bridge the distance between leech and human because of the way leech–human relationships were understood by nineteenth-century medical culture (which we return to in the next chapter). Modern biology, in contrast, ascribes leeches only the most basic of sensory papillae. These take the form of small projected discs found on each segment of the body, thought to be capable of detecting changes in light and temperature as well as vibrations and chemical 'smell'. As limited as their sensory organs appear, leeches are well able to navigate their worlds. On land or water, leeches are known to be incredibly sensitive to movement, temperature and other, presently unknown, indicators given off by potential hosts. When hungry, predatory and bloodsucking leeches have little trouble locating a meal.

Despite having adapted to diverse ecological niches, the many species of leech share a remarkably uniform anatomy. The smallest leeches are 1 cm in length; more common is 2–5 cm, whilst some (such as *Hirudo medicinalis*) grow up to 12 cm. The giant Amazon leech, *Haementeria ghilianii*, is the largest, at 30 cm long.

Unlike all other annelids, where variety between species is the norm, the exterior body of every leech species exhibits precisely 33 segments. All leeches are hermaphrodites, possessing both male and female genitalia, but unlike other annelids they reproduce sexually, through copulation, for leeches are not simultaneous hermaphrodites. Rather, leeches are 'protandric', meaning their male genitalia mature first, whilst their female genitalia emerge later in life. In human culture, hermaphroditism is widely considered abnormal, because it challenges the male/female binary upon which so much of our identity rests. Leech sexuality, in contrast, alternates through life and is temporally determined. In this way, as in so many others, it must be a curious thing to be a leech.

Most leech species breed annually or biannually in spring or summer. When two leeches mate, their bodies become entwined. Often they grasp one another tightly using their sensitive anterior suckers as though embraced in a long, passionate kiss. Beyond this, when it comes to reproduction, leeches differ according to species. Larger species and those adapted to land possess a penis, which is inserted into the female genitalia, depositing sperm. However, many species do not possess a penis but project their sperm directly into the body of their partner by inserting a capsule containing the sperm (a 'spermatophore'). This hypodermic insemination occurs at or near the female genitalia. Once injected, the sperm is released, to make its way through the host's body to the ovipore (the leech's equivalent of ovaries) where fertilization takes place. This mode of reproduction is common to many creatures and is by no means as painful or distasteful as it might sound. Nevertheless, in the late twentieth-century human imagination, such sexual practices have become synonymous with horror. In films such as David Cronenberg's *Rabid* (1977), the dietary and sexual practices of leeches merge to emphasize the

A pair of Malaysian tiger leeches entwined in a mating dance.

darker connotations of their association with parasitism. But leeches need not be seen as horrific or sexually violent, nor should their passion for blood be considered monstrous.

In their own way, leeches care and provide for their young as any human parent might. Leeches lay their eggs within nutrient-rich cocoons which, depending on the species, are carefully attached to submerged objects such as rocks or plant life, or buried in damp soil. The Piscicolidae (fish leeches) often provide their young with a ready meal on hatching by attaching their cocoons direct to a host. Though uncommon, some leech species care directly for their young. Glossiphoniidae, for instance, possess broad, flat bodies to carry their fertilized eggs on the underbelly. On hatching, the young make their way to the front of the parental body, where they wait to be carried to their first meal. Leeches may lack many things, but not a maternal instinct.

Some species of leech, such as this turtle leech, nurture and carry their young for a period of time until they mature and become independent.

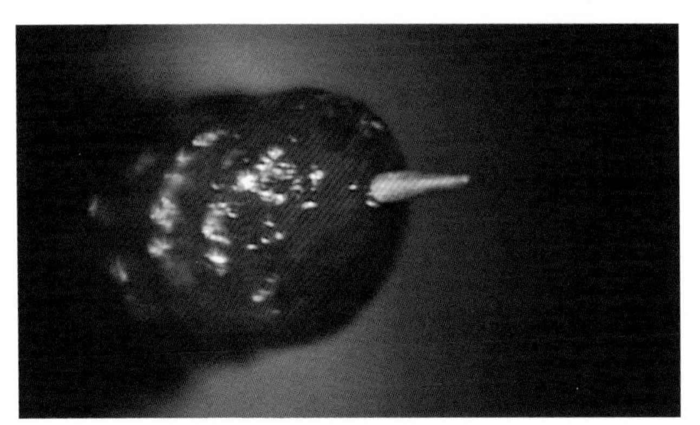

David Cronenberg's *Rabid* (1977) features a leech-inspired 'monstrous' woman who feeds with a leech-like proboscis.

Leeches are born in cocoons consisting of a spongelike mucous material that can contain up to 20 young.

LEECH HABITATS

Leeches live in virtually every corner of the world. Although more frequently encountered in temperate or Arctic seas, marine leeches can be found in tropical waters where they feed on sharks, skates and rays. Piscicolidae are the only leeches to have developed gills, giving them a novel bodily shape as well as a unique means of breathing (other leeches exchange gas through their body surface). Freshwater fish leeches have long been recognized in human culture. In 1451, for instance, the Bishop of Lausanne was forced to expel an 'immense number of enormous blood suckers' from a Swiss lake as they were devastating salmon stocks, an important source of food on fast days.[8]

With the exception of Antarctica, leeches have adapted to freshwater habitats on every continent. Many species have global ranges, though some are confined to specific areas. The Americobdellidae and Cylicobdellidae, for example, are local to South America. Leeches living in Europe, the Americas and Africa are often found submerged in freshwater pools, rivers, swamps or marshlands, or buried in mud or under rocks, unless on the hunt for a meal. In general, freshwater leeches prefer quiet or slow-moving water such as pools and wetlands, as opposed to fast-running rivers. In such places, their numbers can be vast: one report found a single square metre of Illinois water that contained over 10,000 leeches.

Though rarer than their water-dwelling kin, terrestrial leeches can be found in perpetually moist, usually tropical, climates,

Leeches swimming in the water, from *Trousset Encyclopedia* (1886–91).

Leeches from ancient China, *Cheng Ching Shih Cheng Lei Pei Yung Pen Ts'ao* (Ming edition).

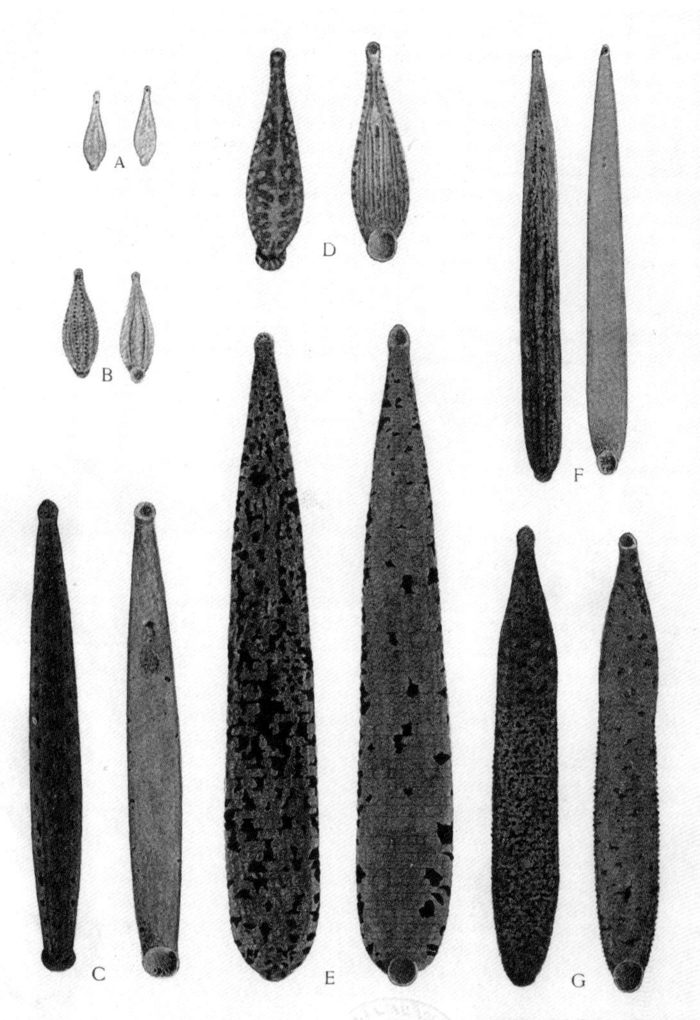

A--Glossiphonia stagnalis
B--Glossiphonia complanata
C--Macrobdella decora
D--Placobdella parasitica
E--Haemopis grandis
F--Erpobdella punctata
G--Haemopis marmoratis

principally Madagascar, India, and across Southeast Asia to the west Pacific and Australia. In humid climates, leeches that have adapted to land are found in trees and on foliage, whilst in drier environments they tend to stay closer to the ground and to moisture. In unusually dry weather, land leeches are known to burrow into the soil like their earthworm cousins where their bodies contract, their skin dries and their suckers become almost invisible. In this state they can survive for many months without food or water, springing back to their leechy selves within minutes of the return of moisture.

Illustration of different species of leeches, from Henry F. Nachtrieb, *The Leeches of Minnesota* (1912).

Behaviourally, the leech exists in two extremes, sedate or active. Leeches can survive long periods of time without feeding,

The four-armed Indian healing god Dhanvantari holding a conch, a disc, a vase containing the nectar of immortality, and leeches for bloodletting. Dhanvantari is the symbol of Ayurvedic medicine, an ancient Indian healing practice in which leeching has always played a central role.

lounging at the bottom of pools, in tropical rainforests or sometimes buried in mud. A hungry leech, however, will be constantly on the move, exploring its environment on the search for potential hosts. Warm-blooded animals, therefore, serve as another leech habitat. By partaking in the most intimate of touches and exchanging the most personal of fluids, leeches transgress bodily borders when feeding. The point of transgression is the blood-sucking leech's anterior sucker, containing the mouth and the teeth, the latter capable of piercing the host's body. For the leech, therefore, to feed is to kiss (of which more momentarily).

For zoologists, the mouth of the leech is less an opportunity for sensuous encounter than yet another way to make sense of these beguiling creatures. The functional and physical anatomy of the leech mouth provides a basis on which to group species into two broad classes: jawed (Arhynchobdellida) and jawless (Rhynchobdellida). The former can be further subdivided between those with (Gnathobdellida) and without (Pharyngobdellida) teeth. Though many leeches are blood-feeders, about a quarter are carnivores. The Pharyngobdellida, lacking teeth, feed on small invertebrates such as snails, insect larvae and even small worms, swallowed whole. Amongst blood-feeders, there is much variety in mouth anatomy. *Hirudo medicinalis* (the medical leech), for one, possesses three jaws, each equipped with many tiny yet sharp teeth capable of piercing even the toughest of skins. Blood can then be drawn from the host into the leech through powerful muscular suction. The bite of this three-jawed leech leaves a distinctive Y-shaped mark within a circle, which some liken to the logo of Mercedes-Benz (but which also bears a passing resemblance to the international symbol for biohazard). Perhaps most distant to human understanding is the jawless Rhynchobdellida. These species possess a proboscis, an elongated muscular appendage which, when projected from the mouth, pierces the skin of

Upright terrestrial leech poised on the anterior sucker. The sucker at the head end waves in the air seeking pasing animals. In 1946 George Stifler Seagrave described leeches which, as he moved, would 'sway like a compass needle and keep pointed right at you'.

the host allowing the leech to draw blood as if through a straw. Such is the stuff of nightmares. Even the most audacious human weavers of horror have rarely dared to transpose this mode of feeding from the leech to the humanoid monster. The vampire, that most humanized incarnation of the leech, feeds as does the comparatively civilized *H. medicinalis*. Only in stories that some would argue should never have been told has proboscis been embodied within the human. In the deeply troubling exploitation film *Blood Sucking Freaks* (1976), a male doctor removes the teeth of his female victim, then pierces her skull and sucks out her brains through a straw, horrifically merging and distorting leech behaviour and the very worst of human nature and culture.

Whilst the peculiarity of leech behaviour has stimulated much horrific fiction, it has also presented difficult questions for

Cross-section of *H. medicinalis*.

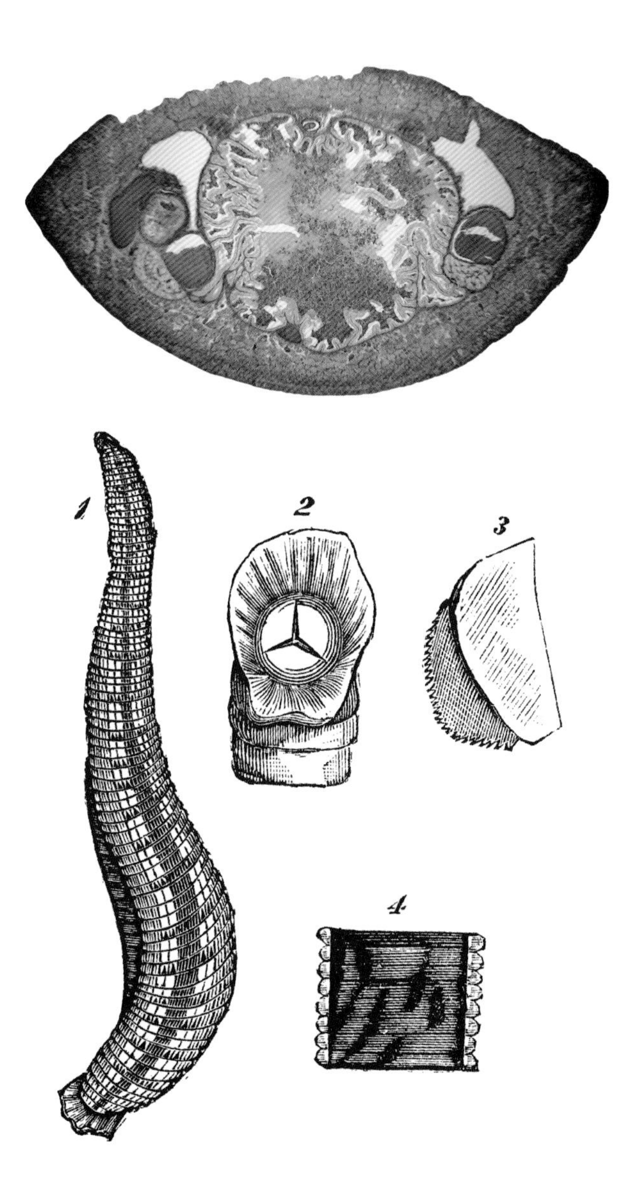

19th-century engraving of a European medicinal leech, *Trousset Encyclopedia* (1886–91).

the zoologist. Many Rhynchobdellida, for example, can dramatically change their outer body colour. Unusually, this behaviour has little to do with camouflage; regardless of their surroundings, the creatures become much darker when exposed to light, yet lose their colour when placed in darkness. The speed at which these colour shifts occur is itself a physiological puzzle. It is tempting, perhaps, to suppose these leeches might change their outer colour at will. Some have suggested that what we see as colour change may in fact be a means by which leeches orientate themselves within the world: controlling the skin's sensitivity to light may allow leeches to detect and follow shadows falling upon them, thereby locating potential hosts. Others have suggested that this rapid transformation in colour may protect a leech's vulnerable body from direct sunlight, which most leeches are known to avoid at all costs.[9]

Leeches' habit of avoiding prolonged exposure to sunlight is difficult for many humans to understand. We relish the sun, risking its dangers so as to obtain the most minor change in skin colour. Despite medical warnings against overexposure, we cannot help

H. medicinalis at work.

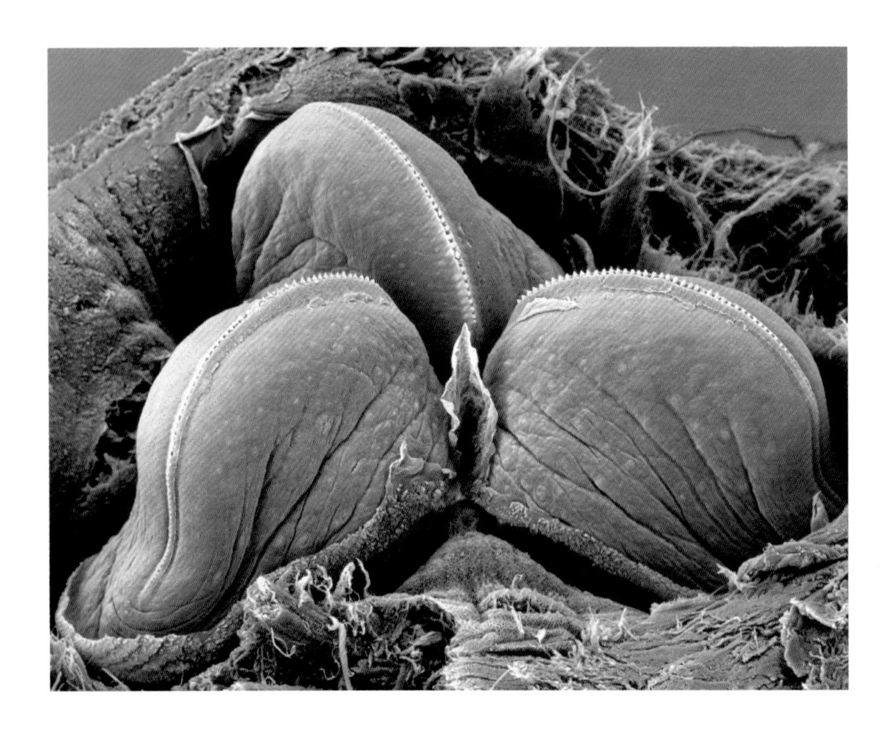

Magnified image of the mouth of a medicinal leech, illustrating the three jaws containing hundreds of tiny razor-sharp chitinous teeth. These puncture and peel back the host's skin, creating a Y-shaped wound through which the leech feeds.

but be suspicious of those who hide from the light. In the shadows, we fear monsters. In our cultural imagination, aversion to light has itself become monstrous, as in the figure of the vampire. Somewhat unfortunately for the leech, this is not the only behaviour the two creatures share.

BLOOD

Feeding upon the blood of others is so deeply associated with the leech that it has become almost synonymous with their very being. Blood is part of the natural and cultural identity of leeches. This association is neither undeserved nor unwarranted. Recognizing

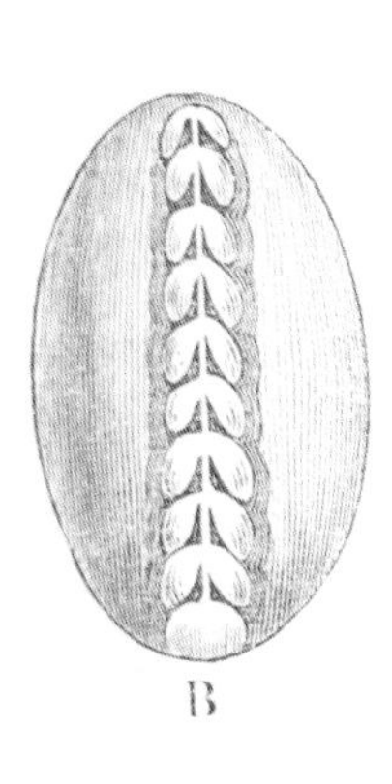

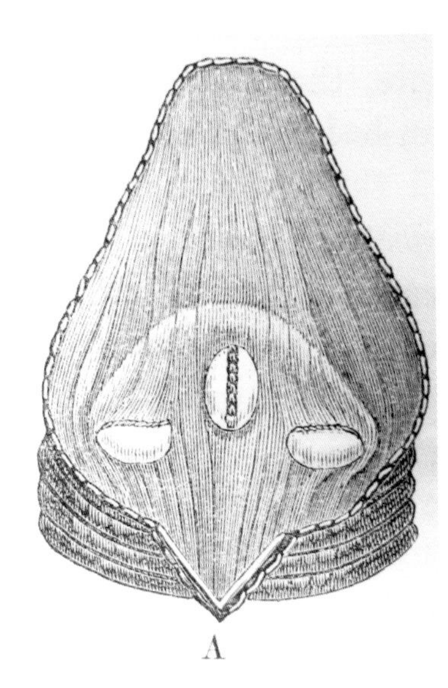

Leech mouth (A) and tooth (B) from Thomas Rymer Jones, *General Outline of the Organization of the Animal Kingdom, and Manual of Comparative Anatomy* (1871).

that nature and culture are less distinct categories than symbiotic systems of meaning can help to explain why, for instance, Cuvier's definition of the leech as a 'red-blooded worm' was so easily accepted. The majority of leeches are sanguivorous (blood-feeding). Being capable of ingesting immense amounts of blood proportional to their body size, their digestive process is comparatively slow. Though leeches eat irregularly, when they do, they feast. The terrestrial *Haemadipsa picta*, or tiger leech, found in Malaysia, is by no means unusual in increasing its body weight tenfold when feeding. By contrast, the European *Hirudo* are almost civilized in their appetites, drinking only two to three times their weight in blood. It is little surprise that following such a meal leeches become sedate, resting and digesting for long periods. For instance, *H. medicinalis* can take over 200 days to digest a meal, requiring only one or two meals per year. For centuries, the capacity of leeches to survive periods of over a year without feeding has transfixed human observers. It is for this reason that blood feeding has become associated with extended life spans, bordering upon immortality. The vampire has borrowed much from the leech.

However, this does imply leeches are monstrous. We need not view blood feeding as parasitical and therefore negative. Many leeches, for example the *Hirudo,* have evolved a reciprocal relationship with mammals akin to companionship. Such leeches are carefully adapted to draw appropriate quantities of blood, often without causing any lasting harm. Using their razor-sharp teeth, leeches effortlessly open the body of their host, frequently going unnoticed. In return for their meal, leeches exchange complex biochemicals including some that save their hosts from feeling painful sensations. Mild anaesthetisation is of mutual benefit to leech and host. Leeches have also developed a powerful biochemical that prevents blood coagulation. Once a leech has completed its meal, blood can flow from the bite for an hour or more. Although not obviously advantageous, this unique ability has been used in human medical practice to enhance the healing capacities of bloodletting. In the nineteenth century, continued blood flow was considered beneficial, since leeches were used to remove quantities of blood from the patient. The modern biomedical sciences, meanwhile, have isolated the substance responsible from the salivary glands of leeches and put it to use as a means to prevent and remove blood clots. In the twenty-first century, leeches are increasingly recognized as nothing less than 'living pharmacies', producing unique biochemicals which may have important medical utility. From this wider perspective it is hard to claim that the leech occupies a parasitical relationship within human cultures. Would it not be fairer to imagine leeches, and their relationships to others, more positively?

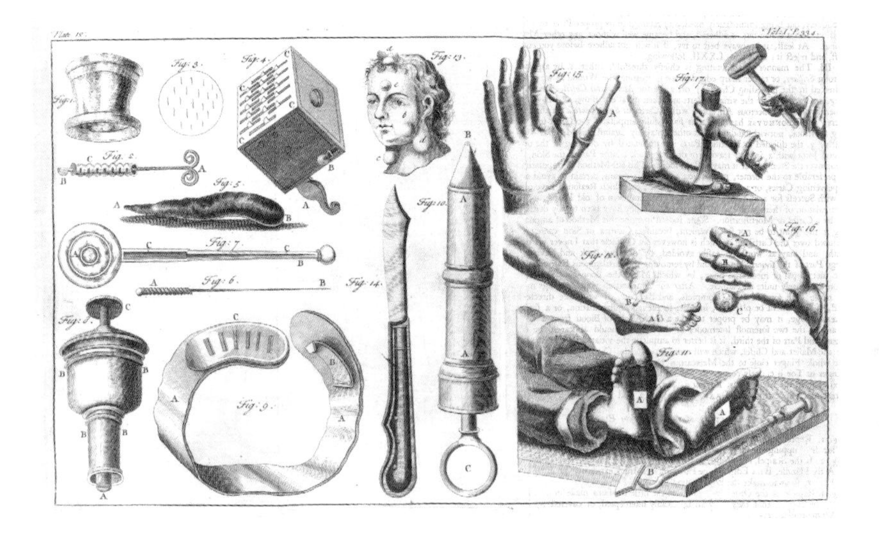

The medicinal leech takes its place in the surgeon's tool kit, in Lorenz Heister, *A General System of Surgery* (1743).

A sanguivorous life, after all, cannot be lived alone. As remarkable as leech biochemistry might appear, they lack an astonishing array of biochemical secretions necessary to survive on a diet of blood alone. Their digestive systems produce no amylase, an enzyme required to break down sugar. Nor can leeches produce lipase, an enzyme that breaks down fats, or endopeptidase, which helps break down amino acids. Without these, it is impossible to transform blood into the nutrients necessary for life. Consequently, leeches depend upon a range of internal life forms that call the leech their home. A leech not only needs a host, therefore, but must be a host. Within *H. medicinalis*, bacteria, such as *Aeromonas hydrophila*, labour hard to break down proteins, fats and carbohydrates from the blood ingested by the leech. Without hardworking gut bacteria, leeches would be unable to digest their bloody meals. When feeding, leeches feed not only themselves but also their internal fellow workers, which flourish once their host begins feeding upon a

In the *Magic: The Gathering* card game, 'Leeches' both harm and heal the player.

host. These bacteria are also thought to be responsible for producing vitamins and other substances necessary for the leech to live. Different species possess different bacteria, for instance *H. verbana* is host to *Aeromonas veronii*. Understanding these relationships 'all the way down' is critical when using leeches in medicine. Leeches can regurgitate their bacteria into the host when feeding, particularly if forcibly removed. If infection occurs following hirudotherapy, treatment may be hindered if one has mistaken the species of leech, for one will look for the wrong bacteria. Just as many species of leech have perfected the art of living on and living with their hosts, so too have others learned to live with and in the leech. There are wheels within wheels, and perhaps we should not be too quick to label some relationships parasitical and others symbiotic. Familiarity, however, is critical if our shared relationships are to be productive.

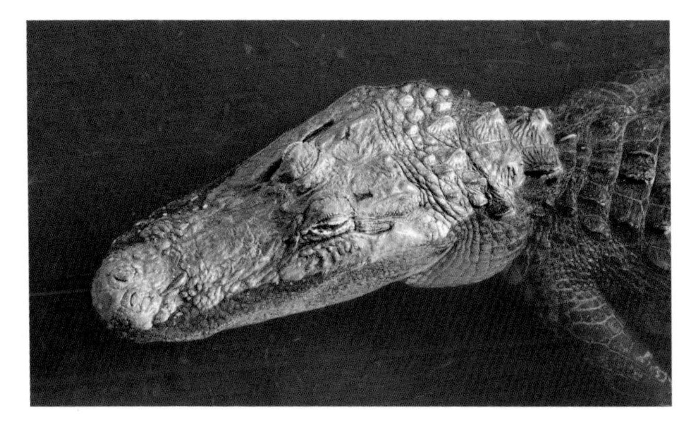

American alligator (*Alligator Mississippiensis*) from the Florida Everglades with a leech attached to the snout.

The evolutionary relationship between carnivorous and sanguivorous leeches is, as is so much else with leeches, a family secret. Traditionally, taxonomists believed carnivorous leeches to be evolutionary predecessors of their blood-feeding kin. However, modern DNA analysis has recently revealed that all leeches share a single sanguivorous ancestor, and moreover that blood-feeding has been given up in favour of a carnivorous diet on at least five independent occasions.[10] These shifts in scientific understanding over time mirror wider changes in the cultural positioning of leeches in human society. In the nineteenth century, blood-feeding leeches were widely considered to be civilized whilst carniverous leeches were thought barbaric. This reflected the comparatively privileged place of blood-drinking leeches within medical practices at the time. Twenty-first-century readers, whose cultural imagination has been shaped to see blood drinking as monstrous, might struggle with such a thought. But no more so than a nineteenth-century reader would struggle with the idea of blood-feeding leeches being the monstrous predecessors of the comparatively more evolved carnivorous species (such as the horse leech, *Haemopis sanguisuga*).

Many leeches possess remarkable colouring.

Today, leeches are mysterious and, for many, repulsive creatures. Yet those who have come to know them intimately recognize their beauty. Mark E. Siddall, curator of Annelida at the American Museum of Natural History and a foremost expert on leeches, certainly thinks so:

> In point of fact leeches are remarkably beautiful. The North American medicinal leech, if anyone bothered to look before tearing one away, is decorated with orange and black polka dots on an olive canvas; the European leech of bloodletting lore has intricate ruby and emerald patterns running the course of its body, and all are among the most graceful of swimmers, plying their watery environment with sinusoidal waves of motion as they stealthily track their targets. One whole family of leeches, specializing in turtles and frogs, exhibits as much or more parental care

as any bird; not only brooding over their young while they hatch, but carrying their offspring to their first blood meal, ensuring a head-start on life.[11]

Perhaps there is more to be enjoyed from leeches than our initial distaste would suggest. It is said that on his deathbed in 1832, Cuvier stopped his nurse as she applied medicinal leeches to his body, not out of revulsion but to remind her, with some pride, that it was he who had discovered that leeches possessed red blood.[12] Why might that have been?

2 Medical Leech

The earliest known reference to leeches in human history occurs in a set of ancient Babylonian texts dating from the second millennium BC. Babylonians described a striped bloodsucking worm which, after feeding, became 'thick with blood', leaving a 'sickle'-shaped mark where it bit. This worm was known to pose a particular danger to the unwary. One account, for example, tells of an infant whose eye has been attacked, likely when the child was washed with river water that inadvertently contained a leech. When bitten by wild leeches, Babylonians would have been treated by a healer – part physician, part magician – using mystical incantations, some of which survive today. However, not all references to leeches are negative. Though Babylonian medicine is not traditionally thought to have practised phlebotomy, or medicinal bloodletting, evidence exists to suggest that leeches were used for this purpose.[1] In one text, a leech is described as 'the daughter of Gula', the Babylonian goddess of medicine and healing. In ancient Babylonian culture leeches were a danger to health whilst simultaneously being part of the physician's *materia medica*. Leeches, then, have long been both a horror and a healer.

Leeches also appear in ancient Egyptian culture. Thriving in and around the Nile river, which Egyptians believed to be the source of life itself, it is little surprise that leeches were incorporated into the civilization's healing practices. A wall painting from the tomb of the scribe Userhat (*c.* 1567–1308 BC), for instance, depicts leeches being applied to a patient.[2] References to leeches, often relating to healing, can also be identified in ancient Arabic, Chinese, Persian and Sanskrit literature. In Chinese culture, leeches appear in the *Erh Ya,* a somewhat mysterious encyclopaedia estimated to date from about the third century BC. The Chinese scholar Wang Chong (AD 27–100) described the accidental discovery of the healing properties of leeches: a king, suffering from 'constipation of the blood', on discovering a bloodsucking worm hidden in his salad, swallowed the animal in order to avoid embarrassing those who had prepared his meal. Later, the king found himself cured of his chronic affliction. This was, Wang Chong explains, a happy effect of the leech having drawn blood from the site of illness within the king's body.[3] However, Wang Chong goes on to clarify that the king was wrong to have swallowed a leech, since it allowed the negligence of the chef to go unpunished. In most cases a swallowed leech is highly dangerous, for the small animals will latch on to the throat to feed, expanding slowly to eventually suffocate their host.

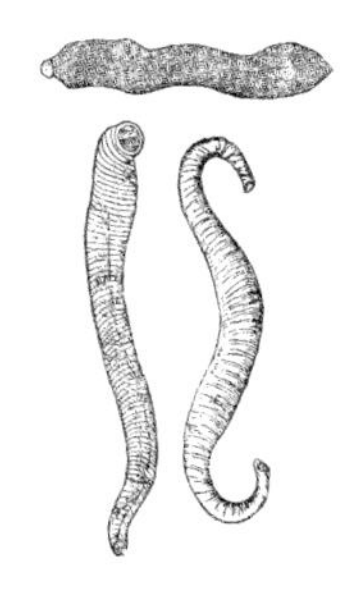

Images of leeches from the ancient Chinese text Erh Ya Yin T'u.

Hirudo medicinalis after feeding.

Detail of the leech mouth and surrounding sucker, from Larive and Fleury, *Dictionary of Words and Things* (1895).

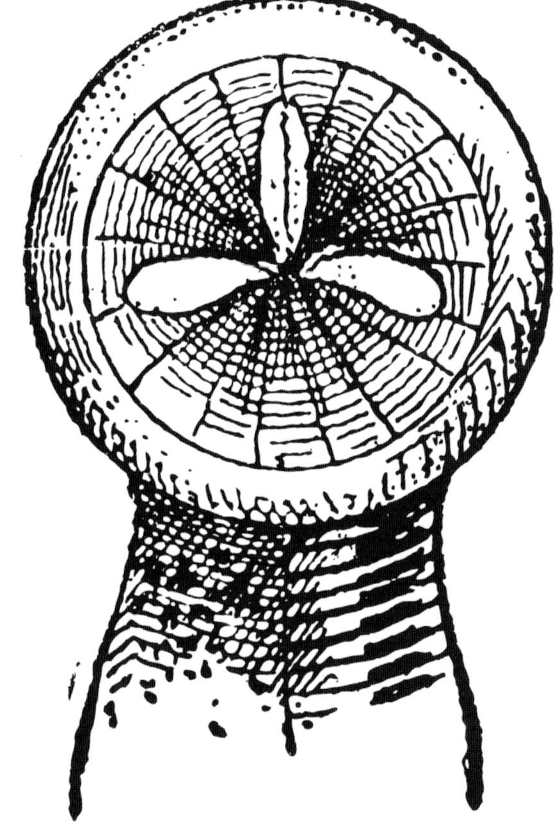

Engraving of a Greek 'Leech Bow Fibula' (brooch) illustrating how leeches have been incorporated into high fashion.

For instance, in 1799 Napoleon's army was struck by a terrible affliction. While marching across Egypt, man after man began to experience

> painful stinging in the posterior . . . frequent cough and expectoration of glairy mucus streaked with blood, and a disposition to vomit . . . to these symptoms succeeded swelling of the throat, frequent haemorrhages, difficulty of swallowing and respiration, pains in the breast . . . perceptible emaciation, loss of sleep and appetite, restlessness, agitation and death.[4]

Napoleon's chief surgeon, Baron Larrey, discovered the men to have a strange black growth deep in their mouths and nasal cavities. One can only imagine his shock, and that of the patient, when Larrey touched the growth only to see it move! The men were infested internally with leeches, easily swallowed when small if drinking water is taken from a leech-filled stream. Now lodged internally and gorged with blood, the leeches had grown to immense sizes. Those leeches that could be reached with forceps were forcibly removed, and the soldiers were then made to drink vinegar in the hope it would encourage those deeper within the body to detach themselves. Many lives were lost to bloody and painful deaths.

It is in ancient Greece that leeches were first systematically incorporated into medical practice as a calming alternative to other forms of bloodletting. Some translations of the *Iliad*, Homer's eighth-century BC epic poem set during the Trojan War, describe how:

> a leech is worth many other men for the cutting out
> Of arrows and the spreading of soothing medicaments.[5]

By the second century AD, the Greek physician Antyllus was using leeches for all manner of ailments, particularly where patients feared the lancet.[6] Leeches were also prominent in Roman medical culture. Pliny the Elder (AD 23–79) recommended the application of leeches for gout and other diseases. He also reported the dangers of wild leeches, which he noted had 'begun to be commonly called a bloodsucker', describing elephants that had experienced extreme pain on swallowing the animals.[7] Despite neither Hippocrates (*c.* 460–370 BC) nor Galen (*c.* AD 129–216) advocating the use of leeches, it was the humoral theory commonly ascribed to those physicians that sustained the leech's place in Western medicine for centuries. Humoral medicine understood the body to be filled with four cardinal humours: blood, yellow bile, black bile and phlegm. Each corresponded to a range of phenomena, including the seasons (spring, summer, autumn and winter) and human life cycle (childhood, youth, adulthood, old age). Disease was not considered an entity in itself but rather the result of disturbed humours; health, correspondingly, was a matter of maintaining humours in balance. Phlebotomy formed one of the primary means of 'evacuation' by which physicians could intervene to release excesses and maintain the humours in correct proportion. Bloodletting was not merely a cure, it was also preventative. As blood was associated with the season of spring, it was common practice to bleed at this time so as to remove excess blood, preventing humoral imbalance.

The Persian polymath Avicenna (980–1037), in *The Canon of Medicine*, unified humoral theory with the use of leeches. He established the rudiments of leech care and their safe handling, explaining which leeches to use and which not, as well as how to recognize an unhealthy leech. Those with black or muddy excrement or whose presence blackens clean water rendering it offensive in smell, and those with a soft mouth, were to be avoided.

These were signs that the leech was unwell, illustrating that physicians not only wanted to use leeches as medical therapies but to do so had to care for the health of their animals. Avicenna's favoured leech possessed a greenish skin with orpiment (brownish-yellow)-coloured longitudinal lines along its body. This is almost certainly a description of *H. medicinalis*, a species whose name owes as much to its privileged position within human medical culture as to its biology. Prior to use, Avicenna recommended, leeches should be squeezed to encourage the ejection of stomach contents and then cleansed, as should the site of application and the physician's hands. Leeches could be encouraged to bite by the application of sugar water or milk to the site, or by puncturing the skin so as to release a drop of blood. A major advantage of the leech over the lancet was that it could be placed directly on any part of the body – though as living animals, leeches, of course, have a mind of their own and often sought to dine at sites more preferable to their own desires. At worst, leeches were known to venture within the body if allowed to stray close to human orifices. In such cases, Avicenna recommended the use of salted water, either swallowed or by enema, as the best way to encourage the animals to vacate their new home.

Once in position leeches were generally allowed to work until they chose to detach themselves, which could be anything from 30 minutes to an hour. Physical removal was thought equally damaging to leech and human and so was discouraged. However, if a patient began to feel faint, physicians could intervene by sprinkling salt, pepper or ashes onto the leech, causing the animal to detach. Suction from the application of a cup to the wound could be used to draw out toxins, whilst the application of quicklime, ashes or ground-up earthenware would help bring bleeding to a halt.[8] *The Canon of Medicine* was used as a standard university text as late as the seventeenth century, giving Avicenna's principles immense longevity. For centuries, elite physicians, who were educated to shun innovation and look only to the past for wisdom, followed Avicenna, establishing *H. medicinalis* as the medical leech of choice.

The holistic nature of humoral theory was difficult to disprove and almost infinite in its narrative flexibility. Moreover, in humoral medicine, interventions could be prescribed with or without symptoms of illness. This meant that the leech could be deployed to treat all manner of complaints. In medieval monastic life, for instance, bleeding extended beyond the boundaries of medical necessity: bloodletting took on religious elements, becoming associated with purification and the sacrifice of Christ. Most abbeys possessed a 'bleeding-house', where regular phlebotomy (known as the 'seyney') could be practiced at stated periods of the year. Most healing at this time was carried out by members of the clergy. However in this case bloodletting was less medical than a quasi-sacred practice finding its place in the religious calendar.[9] Participation in the seyney allowed the relaxation of the strict rules that governed the daily life of monastic orders, particularly their dietary restrictions. Spiritual bloodletting, then, was more than the purifying of body and soul: it was something of a social event. And leeches were at the heart of the party.[10] For this reason, bloodletting was forbidden at certain times of the year (particularly during Lent).

The humoral idea of health as a balance was also woven into the wider web of cultural beliefs and customs of ordinary people. Since leeches were widely available in rural areas and could be used with the bare minimum of knowledge, leeching was easily integrated into the practices of untrained and often illiterate lay medics. In contrast to the lancet, an expensive and in the wrong hands dangerous surgical tool, the leech was much less likely to cause serious damage to the body. Individuals could even treat

de gresse que les braceletz de sa femme luy seruoient d'an-
neaux a ses doigtz, comme les historiens escriuent. Comme

Histoire d'un Roy qui estoit si gras, quil se faisoit tirer sa graisse a-uecques des Sangsues.

en semblable ce grand tirant Denis Heraclet se laissa si-
bien transporter a ses delices qu'il s'habitua en fin de ne
faire

An unnamed king being treated with leeches to cure his obesity, from *Histoires prodigieuses* (1560).

themselves. Freely available in the countryside from pools and lakes, and widely obtainable in the city from apothecaries and local barber surgeons, by the late eighteenth century leeching was established as a popular treatment for illness of both elite and lay persons.

At the turn of the nineteenth century elite medicine passed through a series of significant revolutions that changed it forever. Humoral medicine, though remaining influential late into

Worried about his sexuality, Lord Edmund Blackadder consults a doctor, who suggests leech therapy to commence as soon as possible in the BBC TV series *Blackadder II* (1986).

the nineteenth century, declined in popularity as new rational approaches to medicine became fashionable. Driving these transformations was a newly formalized institution, the hospital, wherein disease was reimagined. Illness was no longer seen as an imbalance of humours; instead it became an element in itself, to be identified, named and studied. Medical teaching no longer privileged the past as its source of learning. Dusty old books were left on the shelf as the present became the source of new knowledge. Within hospital medicine, physicians gained new ways to access, study and catalogue illness, drawing connections between patient experience, symptom and biological sign. Medicine was becoming 'modern'. Curiously, as result of these fundamental medical revolutions, driven by a modernizing zeal, the leech, that ancient medical companion, became more popular than ever.

Within humoral medicine, phlebotomy had been one amongst many treatments. In the early nineteenth century, however, it became *the* treatment of choice. Advocates of bloodletting emerged across the world but it was in France that the leech became the symbol of the new medicine. Following the French Revolution, the very fabric of the nation was rebuilt according to rational and, above all, modern principles. Modern medicine, after all, should be revolutionary! The future of medicine, or so it was thought at the time, lay with François-Joseph-Victor Broussais (1772–1838). A liberal and a revolutionist, Broussais was gifted with oratory skill, a certain flamboyance and not an ounce of self-doubt. A new medical messiah had arrived, with his own system of medicine opposed to all others. In his bombastic lectures in Paris, attracting large crowds through the opening decades of the nineteenth century, Broussais, in an age of medical innovation, presented his medical system in such a way that the past

In this engraving of *c.* 1832 by J. J. Grandville, three anthropomorphized ' medical leeches' prescribe a treatment for their grasshopper patient based on Broussais' understanding of medicine.

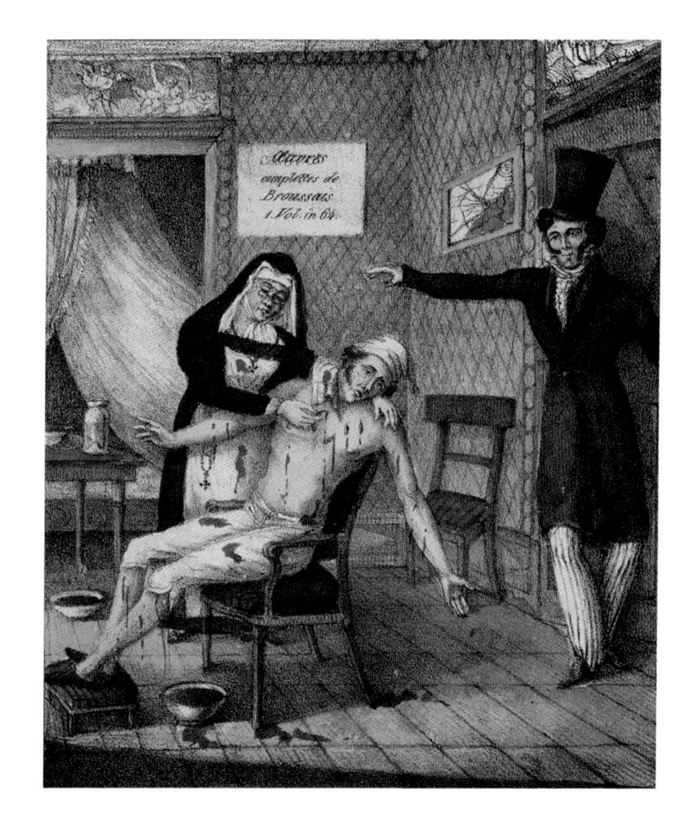

'Broussais' System'. In this print Broussais is directing a nurse to continue leeching an already pallid, blood-besmeared patient. The caption reads: 'But I don't have another drop of blood in my veins!' 'It doesn't matter: another 50 leeches!'

decade of revolutionary medicine began to look centuries old. Disease, Broussais held, was to be understood physiologically, as the result of inflammation, itself caused by over-stimulation.[11] He dismissed the relatively recent idea of individual diseases, such as cancer, smallpox, syphilis and tuberculosis. All, he proclaimed, were the result of inflammation, the root of which was the stomach. The fate of the stomach was to be irritated, and its irritability would spread to other organs. All diseases, therefore, had one cause and one cure: the application of leeches.[12] Had his popularity endured, today we would not recognize a doctor by the stethoscope about their neck. Rather, the potent symbol of medicine would be the medicinal leech.

In remaking the leech as the cutting-edge cure for all disease, Broussais was preaching to a ready audience. As a result of the revolutionary wars, leeching was enjoying heightened popularity across France. Civilian surgeons had been absorbed into the military leaving few with the skills to perform bloodletting via the lancet. Broussais' brilliance was to present an entirely new system of medicine that sounded modern yet was grounded in a simple, familiar and apparently safe therapy. For the patient, leeching was a vastly preferable treatment to recently developed rival medical therapies. The latter, consisting of a mixture of drugs, purgatives, emetics and other tonics, entailed a far from

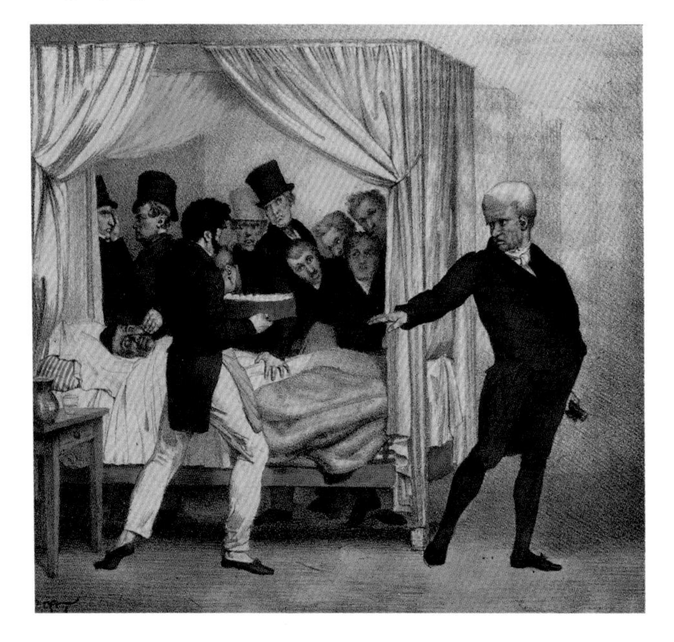

A doctor prescribes another 90 leeches in this French print from 1827.

pleasant experience. Moreover, many of these interventions, all of which Broussais forbade, were not just unpleasant but dangerous. Leeches may or may not have been able to cure, but neither were they likely to be fatal. Thus Broussais, in effect, gained great advantage over his rivals because his method was the one less likely to kill his patients!

Much of the medical culture Broussais dismissed had only itself emerged in the 1790s. Moreover, his own therapy, leeching, was ancient. Nevertheless, in a triumph of flair over fact, Broussais had convinced much of France that his was the medicine of the future. Henceforth there was one cure for all: the leech. And all meant *all*: veterinary medicine, too, was to be reformed. Broussais leeched his fighting cocks weekly to keep them in top fighting condition. He was most puzzled when they lost all vitality, their crests becoming colourless and limp, and they subsequently failed to excel in the fight. He also leeched himself, frequently self-prescribing 50 to 60 leeches to cure his indigestion.[13] By the 1820s, Broussais had not only established the leech's place at the heart of modern medicine but also embedded them within French culture more generally. At the height of leech mania, Parisian salons were filled with chic women wearing dresses adorned with embroidered leeches and ribbons styled like chains of leeches, a fashion known as *les robes à la Broussais*.[14] The French appetite for leeches is startling to modern eyes and revealed in the following international trade figures:

YEAR	NUMBER IMPORTED	VALUE IN FRANCS
1827	33,600,000	1,009,000
1829	44,600,000	1,337,000
1831	36,400,000	1,093,000
1832	57,491,000	1,724,730

1833	41,600,000	1,250,000
1835	19,600,000	677,000
1837	25,800,000	774,000
1839	22,400,000	672,000
1841	17,500,000	524,000
1843	17,600,000	528,000
1844	15,224,671	456,740

Leech imports into France, 1827–43.[15]

Broussais' dramatic success in propelling leeches to the height of medical modernity owed much to his political astuteness.[16] Broussais was careful to remain on good terms with the military, which provided him with patients and students and ensured his medical revolution took hold. His systematic theory of disease and simple method of treatment appealed to army surgeons, who no doubt used many of the leeches imported into France during this period. Outside France, however, Broussais' politics would have been a hindrance rather than asset. Had he earned the nickname 'le vampire de la médecine' because of his aggressive and excessive use of leeches?[17] Or did it imply involvement with brutal French revolutionary politics? In countries such as Britain, which led the defeat of Napoleon and distrusted everything tainted with French revolutionary spirit, the two possibilities were often blurred. Just as the French Revolution brought violence and terror, Broussais was viewed by outsiders as a ferocious and bloody surgeon. Nevertheless, in the middle decades of the nineteenth century, long after Broussais and his medicine had been discredited, millions of leeches continued to be employed by doctors across the world in what has been called the 'golden age of leeches'.[18]

Even in Britain, a most Francophobe of nations, appetite for leeches grew. Not everybody was easily convinced of the efficacy of leeching, however. One sceptic complained in 1825:

Paisley is based on a traditional Indian design thought to have been originally inspired by leeches.

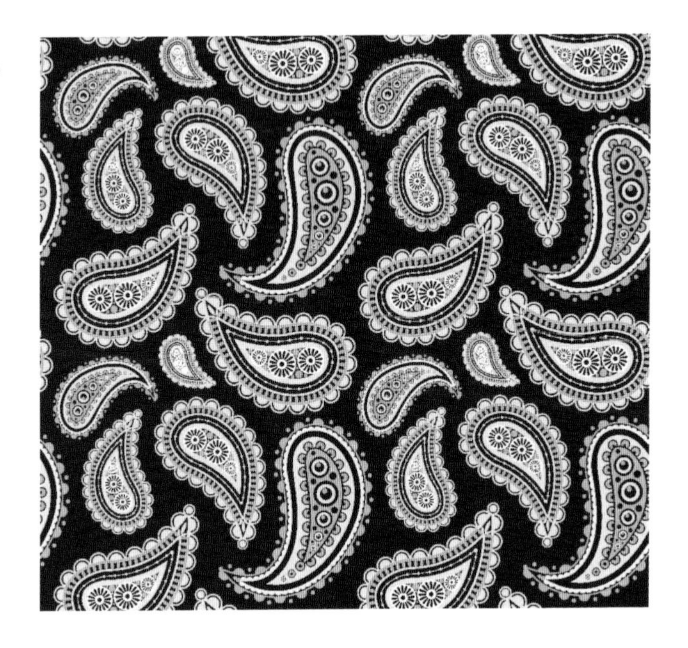

Paris is now a daily scene of Sangrado murders, as London is fast becoming. A hysterical vomiting occurs in a nervous woman: two hundred and thirty leeches are applied; the patient dies as she ought: because, we suppose, she had Monsieur Broussais, gastro-enterite.

A patient, perhaps a delicate girl, is seized by . . . influenza. She is bled, and bled again – becomes worse after each, dies . . . Have we not seen it with our own eyes? . . . It is all the consequence of a fashion, and of nothing else. Physicians seem to have forgotten that there is a nervous system, as well as a sanguineous one . . . what is the cause of all this – this fashion? . . . Simply, because it is easier to follow and copy than to think.[19]

Dr Sangrado was a character immortalized in Alain-René Lesage's novel *Gil Blas* (1715–35), who, expressing the excesses of invasive medical treatment, advocated bleeding of patients as a generic cure for all ills. In the nineteenth century, the name Sangrado was mobilized to critique those physicians who were known to overzealously bleed patients.

Yet comparison to Sangrado was also used to defend leeching. For leech aficionados, it was those who preferred the lancet that risked 'Sangrado murders':

> Within the memory of many of us, bleeding was freely resorted to with a most fool-hardy forgetfulness of the Mosaic sanitary declaration 'the life of all flesh is in the blood thereof' . . . But with our doctors . . . being now much less sanguinary in their curative processes, the lancet is superseded by the leech, whose power of bloodletting fortunately does not extend *ad deliquium animi* – the fainting of the patient from loss of blood – the point barbarously aimed at by doctors of the Sangrado school.[20]

Here the leech is offered as the antidote to brutal venesection, offering a gentler, less extreme therapy than the lancet could ever achieve. Many believed leeches to be a comforting approach to phlebotomy, through virtue of their being a living organism and not just a blunt tool. Unlike the lancet, the leech could be trusted to draw blood from the weak without further taxing their bodies. The leech, therefore, could be friend of the aged, women and children, all of whom were considered too weak in constitution to face the lancet. Leeches were the tools of the physician who cared.

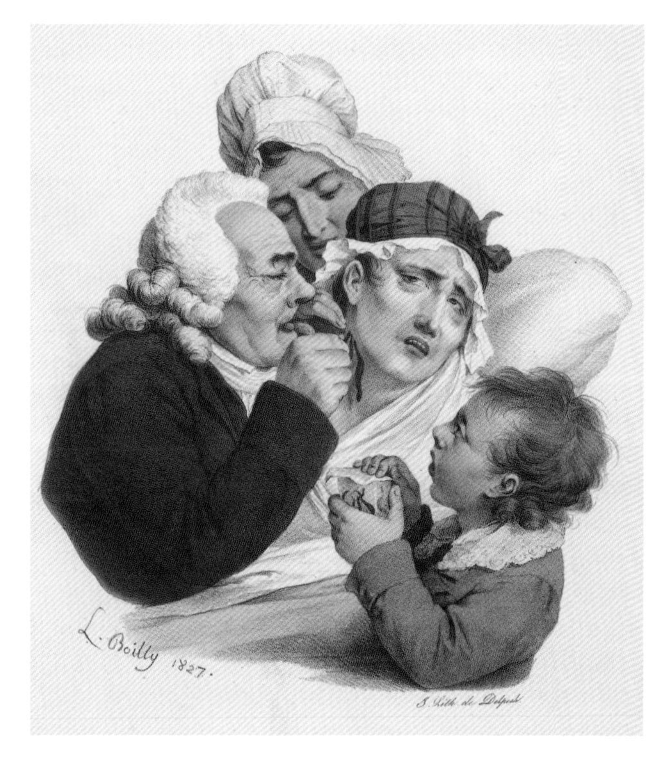

A 19th-century medical practitioner administers leeches to a female patient whilst a young boy holds a jar containing the leeches preparing for work. Louis Boilly after François Séraphin Delpech (1827).

HIRUDO MEDICINALIS

In 1758, in his project to classify the whole of nature, Carl Linnaeus (1707–1778) named one species of leech *Hirudo medicinalis*, literally 'medicinal leech'. Names are important. They tell us who we are and provide a sense of identity. Most of all, they tie nature to culture. Linnaeus named this species of leech not only for its biological nature but also for its place in culture. Following Linnaeus, the medicinal leech grew in stature. Knowledge of the medical leech was formalized in dedicated pamphlets and books towards the end of the eighteenth century. George Horn's

An Entire New Treatise on Leeches wherein the Nature, Properties, and Use of that Most Singular and Valuable Reptile is Most Clearly Set Forth (1798), for example, described their natural history, physiological characteristics and how to identify, locate, house and care for them. In these works *H. medicinalis* obtained an independent identity; defined by colour, length and other physical characteristics, the medicinal leech was increasingly distinguished from other leeches. Whereas in the eighteenth century the horse leech would be recommended for medical use, by the nineteenth *H. medicinalis* and the horse leech had quite independent identities, the former considered a cosmopolitan and companionable healer, the latter a wild and savage monster. Horse leeches were reimagined as ravenous beasts, willing to eat anything and everything – far too dangerous for medicinal use. While they do consume whole animals, often earthworms, shredding their bodies between their two larger than average sets of teeth, horse leeches are not in fact blood feeders. They would have been neither dangerous nor useful in phlebotomy.

In 1816 James Rawlins Johnson differentiated the natures of leeches, concluding none were as aggressive as the horse leech. The appetite of the horse leech, he wrote, drove the creature to devour even their own kind. In a series of experiments, Johnson showed horse leeches to be insatiable cannibals, capable of swallowing *H. medicinalis* whole and willing to fight with others for the privilege.[21] Having introduced a medicinal leech gorged with blood to a jar of around 40 horse leeches, Johnson observed how they seized and dragged their cousin to the bottom of the vessel. The orgy of violence was then obscured as the water turned bloody; the medicinal leech was never seen again.

Accounts such as these did more than demonize the horse leech. They operated as a foil by which the medicinal leech could be seen as moderate in their appetite and civilized in their relations to other creatures. Where horse leeches would never be satiated, drinking blood, eating flesh, even swallowing animals whole, *H. medicinalis* was more refined, dining exclusively and moderately on blood. The horse leech as fiend was quickly absorbed into the public imagination. For example, when a surgeon directed leeches to be applied to a child who subsequently died, a witness reported that the leech had in fact been a horse leech. Widespread public alarm followed, causing a 'violent clamour' to be raised in the neighbourhood.[22] Cultivating the horror of horse leeches helped to protect the expert role of surgeons and physicians. Yes, anybody could take and apply a leech to treat a sickness, but could you be sure you had identified the medicinal leech and not one of their monstrous cousins? Even in the city, where leeches were widely available from apothecaries, one could never be sure. Leeches were sold by weight in large numbers and unscrupulous wholesalers were known, or at least rumoured, to throw in some horse leeches in order to increase their profit margins.

Villainizing horse leeches also helped to integrate *H. medicinalis* within medical culture. The physiology of the horse leech was faster, more demanding, and in having an insatiable appetite was considered unsuitable for bloodletting. By contrast the medicinal leech could be portrayed as calm, docile, gradually removing blood from the body without placing undue strain upon the patient. Thus the more violent the horse leech became, the more civilized a companion the medicinal leech would appear. This also helped leech aficionados assert the superiority of *H. medicinalis* over rival methods of bloodletting such as the lancet, which was portrayed as a violent instrument akin to the horse leech. In this view, the lancet depleted too much blood too rapidly, risking the health of the patient. The mild metabolism of the medicinal leech, however, facilitated a safe and controlled loss of human blood. As the *British Medical Journal* explained:

> Nothing can be safer or more effectual in alleviating the fatal consequences of an overflow of blood, bruises, inflammatory tendency, or other accidents, than the leech, as it can be applied with security to any part of the human body that may be injured without the slightest danger to the patient, when the lancet can only make its incision in the adjacent veins, and that often at great risk to the person of an irritable constitution.[23]

Opening up a vein was seen as ruthless and cruel when compared with leeching. The gentle nature of the medicinal leech also allowed medical treatment to be extended to patients who could not have been treated with the lancet. Leeches increased the power of medicine to intervene not only in cases where a patient was too fragile for the dramatic intervention of the lancet, but to bodies, particularly of women and children, that would never have been bled due to their perceived weaker constitutions. Furthermore, leeches enabled medicine to develop new therapeutic approaches directly targeting the site of illness. Unlike the lancet or any other form of phlebotomy, *H. medicinalis* could be placed internally, extending the frontier of medicine by opening up the internal world of the body. Physicians, themselves emboldened by these new possibilities, encouraged leeches to boldly go where they had never been welcome before.

LEECH AT WORK

From colds to cancer, indigestion to infection, toothache to tuberculosis, the leech was the cure. A common and cripplingly painful male complaint of the time was hernia humoralis, the straining and consequent inflammation of the testicle. This affliction, one surgeon explained:

> demands the application of leeches; and a more convenient part of the body does not present itself for this purpose than the scrotum . . . half a dozen or a dozen leech bites . . . hardly fails of giving great and speedy relief; and the operation should be repeated frequently, until the pain in the part has ceased and the swelling less intense.[24]

As hard as it might be to imagine that swelling in such a sensitive area would reduce in response to the bites of six to twelve leeches, leeching was firmly believed to be effective at the time. Diseases such as hernia humoralis illustrate the way leeches augmented the powers of medical intervention. Unlike the lancet, leeches could be applied directly to the site of the complaint.

However, whilst one *used* a lancet, one *worked with* a leech. As living organisms leeches possessed an agency of their own. When employing a leech, the bloodletting encounter became a collaborative endeavour. Working with leeches required careful negotiation, in the form of interspecies communication, establishing a unique set of relationships between patient, practitioner and leech as medical tool. But how does one communicate with leeches? As physicians grew to know *H. medicinalis* they refined their techniques, coming to see themselves as working within a medical 'team'. Leeches were frequently anthropomorphized, assumed to share human tastes; thus excessive temperatures or repugnant emanations such as human sweat were thought to cause leeches to lose their appetite. Similarly, washing and shaving the skin of a patient was encouraged, since leeches were

Hirudotherapy.

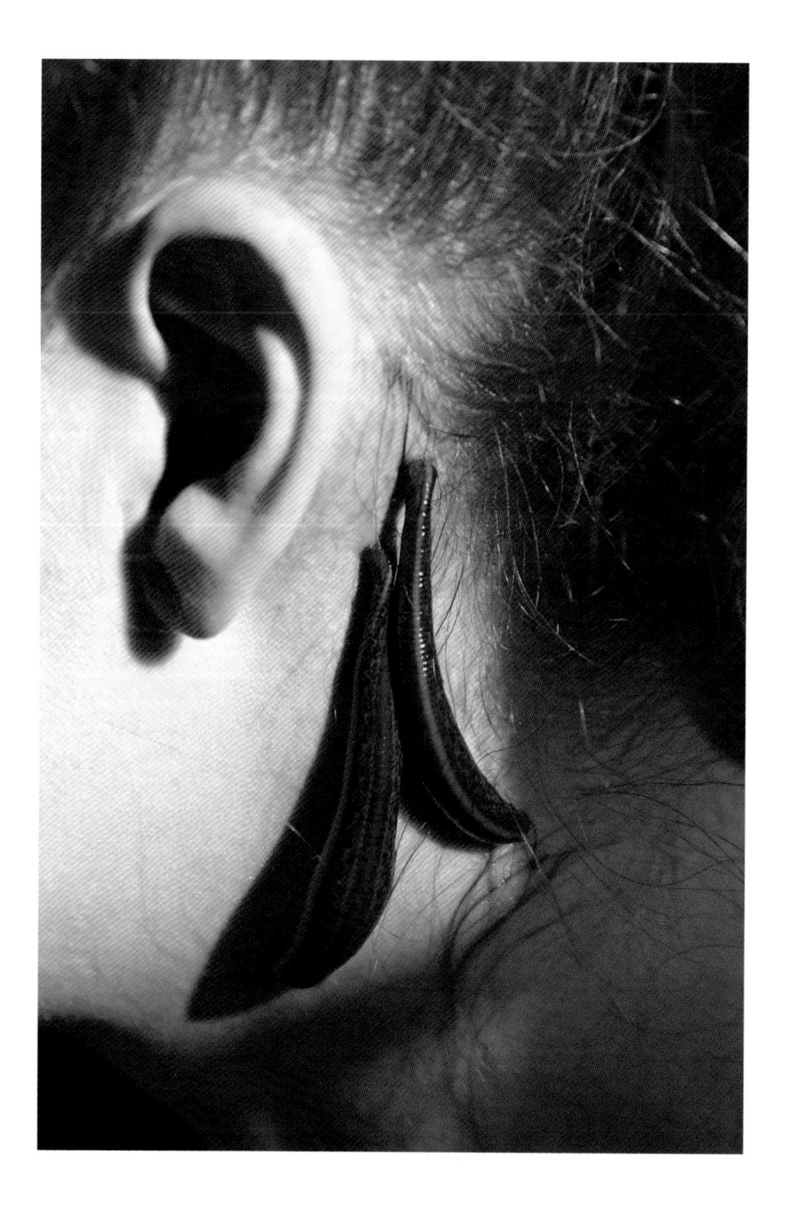

French portable leech jar manufactured by Chevalier, *c.* 1890. The device allowed practitioners to transport leeches pre-loaded in 17 glass tubes, allowing their direct application to patients without any loss of time.

believed to prefer a clean dining environment. Lethargic leeches, a common complaint in the winter months, could be awoken and set to work if doused with wine or beer, for 'leeches drunk will bite until sober'.[25] Alcohol or blood smeared on the skin was used to alert the leeches to their place of work.

The nineteenth century witnessed many new technologies to aid communication between physician and leech. For instance, the practice of placing leeches in a glass, which was then held to the skin until the animal bit, led to the production of specialist 'leech glasses', which came in a range of shapes each designed to allow the physician to apply his fellow worker directly to difficult to reach areas (such as the gums). Here, again, physician-surgeons differentiated themselves from lay practitioners by using specialist and expensive equipment that allowed them to refine and display their medical prowess. Using these thin glass tubes, leeches

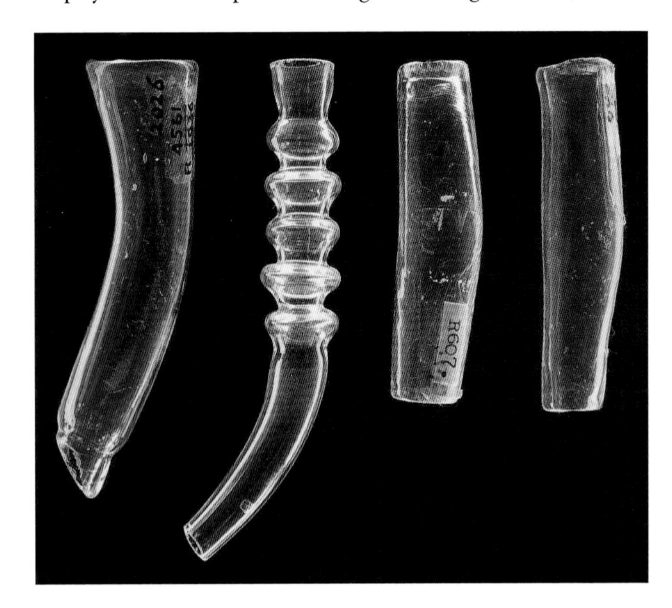

A variety of specialist leech glasses used for placing leeches at the site of treatment.

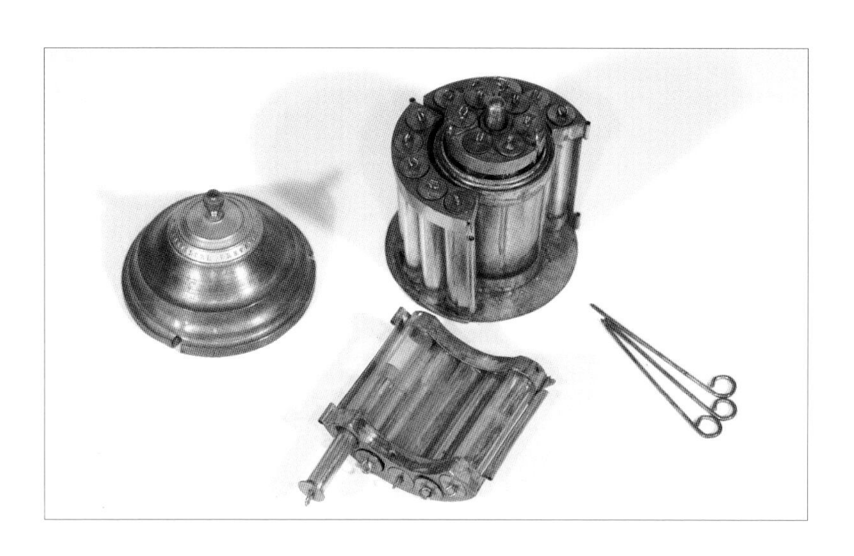

CHEVALIER
PARIS
MONTMARTRE

could be precisely placed in the most difficult to access parts of the body, allowing blood to be drawn from around and within the ears, nose and the mouth, where previously bloodletting had been impossible.

One widely known problem when working with leeches was their tendency, if not watched, to wander from their site of application into human orifices. To guard against leeches crawling into the mouth, nose, ear or anus, a thread could be drawn through the tail end of the animal to secure it. With increased experience, however, some physicians and surgeons sought to adapt what had been a problem into an opportunity. New technical devices were designed to open up human orifices, allowing leeches to be expertly placed within the body. In 1833 Jonathan Osborne, a Dublin-based surgeon, reported a new method for treating inflamed intestines. Hitherto, introducing leeches into the body anally had been a very difficult procedure due to the constrictions of the sphincter. Osborne designed a new instrument by which leeches could be safely transported through the rectum, deep into the body: leeches would be tied to a thin thread to ensure these intrepid medical explorers would not become lost inside the patient. 'I have never used more than four leeches at once', Osborne reported, as this was enough to induce the 'immediate benefit which was to be expected from the direct abstraction of blood from the inflamed surface'.[26] Leeches were also deployed through the mouth to treat bronchitis, laryngitis and to relieve coughs. Inserted vaginally, leeches were thought capable of preventing abortion, easing menstruation and menopause, and treating diseases such as pelvic congestion. The anus presented an inviting opportunity for leeches to treat enteritis, inflamed bowels and problematic prostates. Indeed, so safe was the leech that it was the common practice of one gynaecologist 'to apply a few leeches when the patient is so

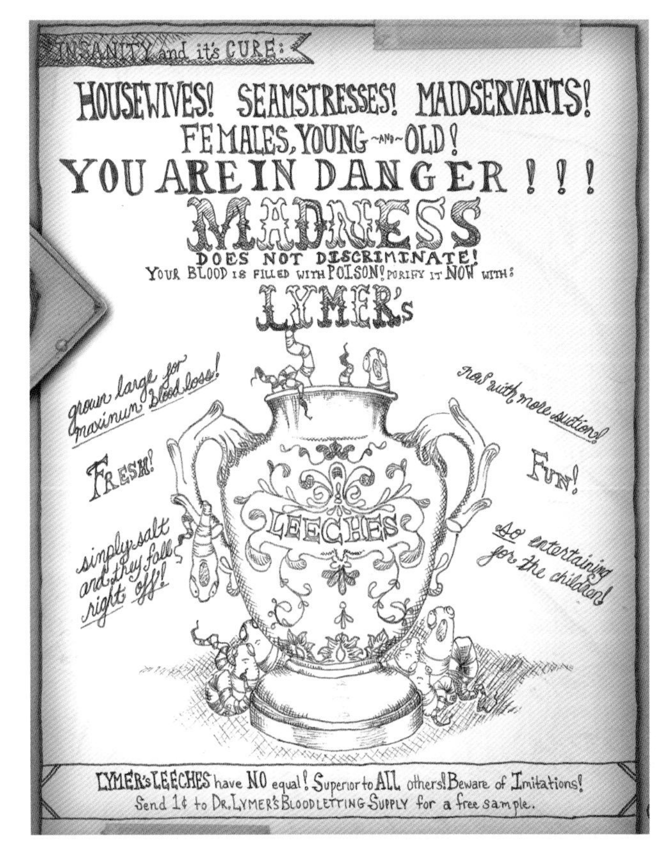

The historical connections between women, mental health, medicine and leeches inform this 21st-century image by singer and artist Emilie Autumn.

weak and anaemic as not to have a drop of blood to spare and with good results'.[27]

These intimate encounters illustrate the degree to which leeching was considered safe, but was the leech a companion animal? Certainly, many who worked with them felt a duty of care towards their leeches. For example, the sprinkling of salt upon leeches was often recommended in the literature of laymen, such as *The*

Household Encyclopaedia; or, Family Dictionary (1859).[28] However, this was considered cruel by those who knew leeches well. Salting a feeding leech caused the abandonment of the meal, whilst salting a fully fed animal would lead to the disgorging of the meal. 'I would ask such inconsiderate persons', one leech enthusiast wrote, 'how they would feel themselves, if, after a hearty dinner, any person was to give them a violent emetic.'[29] But perhaps the best example of a companion relationship comes from Thomas Erskine (1750–1823), the British Lord Chancellor, who famously befriended two leeches. As one of Erskine's acquaintances related:

> He had been blooded by them . . . when he had been taken dangerously ill at Portsmouth; they had saved his life, and he had brought them with him to town, – had ever since kept them in a glass, – had himself every day given them fresh water, and had formed a friendship with them. He said he was sure they both knew him, and were grateful to him. He had given them different names, Home and Cline (the names of two celebrated surgeons), their dispositions being quite different.[30]

3 Capitalist Leech

In 'The Leech Gatherer' (1807), William Wordsworth describes an encounter with a wizened man on a country walk:

> From pond to pond he roamed . . .
> . . . gathering Leeches, far and wide
> He travelled; stirring thus about his feet
> The waters of the pools where they abide.
> 'Once I could meet with them on every side;
> But they have dwindled long by slow decay;
> Yet still I persevere, and find them where I may.'

Once, leeches had been plentiful. Now, they were so scarce that the old man, appearing 'not all alive nor dead', barely had the strength to search them out. The profession of 'leech gathering' had been a longstanding part of the rural economy across the marshy rivers in Lincolnshire, Kent, Yorkshire, Buckinghamshire and across Europe, like the old man, was experiencing a dramatic decline in fortune. The medical demand for leeches had led to their chronic over-harvesting with no regard for the fact that leeches breed slowly. Wordsworth's description of leech gathering as 'hazardous employment' was meant in a dual sense. Whilst the dearth of leeches made seeking them out increasingly taxing, the work itself was also hard on the body. Gatherers stood all day in pools using their bare legs as lures to attract wild leeches. Many shared Wordsworth's concern that the work of harvesting leeches ravaged the gatherers' bodies. In 1798 George Horn, owner of a London apothecary, warned that leech gathering was 'dangerous to a high degree', for the numerous bites would drain the worker and cause alarming inflammations.[1] Others worried that such work involved exposure to 'fogs, mists and foetid vapours', causing 'catarrhs and rheumatisms' amongst other ills.[2] The real danger to gatherers, however, emanated not from rural ponds but from the insatiable urban demand for leeches: this important part of the rural economy was dying with the leech populations themselves, yet leeches had never been more valued. A few years prior, 100 leeches would have cost two shillings and sixpence but by 1808 the price had risen to 30 shillings. Not that this had benefited the leech gatherer, whose body was as exhausted as the ponds that had sustained his now

A man harvests leeches from his legs in this illustration from a 14th-century illuminated manuscript.

Leech gatherer fishing in a marsh in Brittany, France, from *L'Illustration Journal Universel* (1857).

dying trade. Exactly who had drained the life of who Wordsworth leaves undecided; both suffered from the transformation of leeches into commodities traded in a global economy. As the nineteenth century progressed, leeches became both the victim and symbol of capitalism.

LEECH GATHERING

Leech gathering had for some time been an important source of rural income and traditionally worked within nature's limits. Leech gatherers understood, for example, the importance of returning smaller leeches to the rivulets, ponds and ditches whence they came, in order that there would be good fishing on their next visit. Due to escalating demand for leeches, however, new intensive methods of gathering emerged that ignored such traditions. Now, rather than using the human body, pools were disturbed so as to stir curious leeches to the surface, and then dragged with a fine net. This method allowed large numbers of leeches to be caught in one go but left few behind to breed. Only in later years was overfishing recognized to have decimated the numbers of *H. medicinalis* across the world.

With natural stocks declining and demand increasing, British merchants travelled across Europe in search of fresh supplies, transforming leeches into a global commodity. In France, for instance, foreign and domestic demand created a leech trade that quickly became an important branch of commerce. Between 1814 and 1834 the price of French medicinal leeches soared from five francs to over 100 francs per thousand.[3] This was driven largely by the British thirst for leeches. By 1819, British surgeons were increasingly concerned by the sustainability of supplies, having become alarmed that 'we already employ *one hundred foreign leeches* to *every individual* British one'.[4] Merchants consequently built an ever larger network of trade routes by which leeches travelled from Hungarian, German, Silesian and Polish towns to the great cities of Britain and France.

Pewter case used by medical practitioners to transport leeches on home visits.

Urban apothecaries who could be relied upon to maintain a regular stock of leeches were viewed with high regard, since a great deal of labour was required to achieve such a feat. As one never knew when leeches from a particular supplier, region or country might become scarce, sourcing leeches required consistent attention. Overfishing here, or a hot summer there, could devastate leech stocks. Furthermore, governments, concerned with supplying national needs, began to restrict or forbid exportation at times of scarcity. Transporting leeches was no easy task. For short journeys merchants used tightly packed linen bags, which they carried so as to prevent the animals being shaken too much during transit. Curiously, should a thunderstorm approach, merchants would hurry to thrust the bag into the closest pool, marsh or water they could find. Though the reasoning for this is unclear, it is indicative of the close relationship leeches were thought to have with the climate. For extended journeys traders

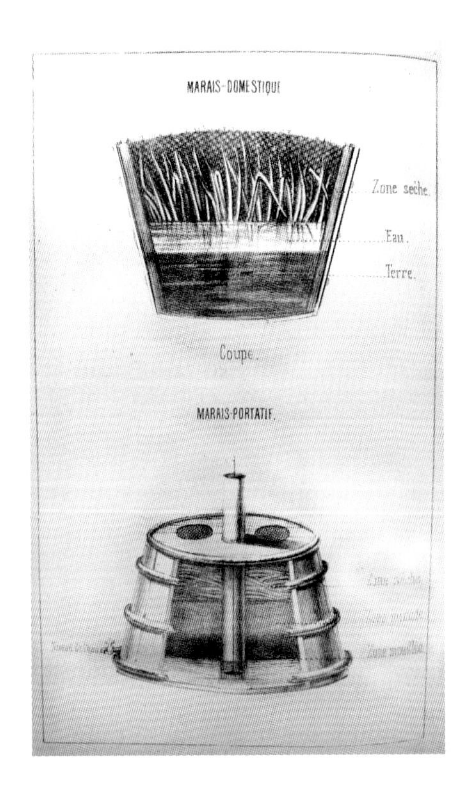

An artificial swamp for the long-distance transportation of leeches, *c.* 1855.

used wooden vessels, carefully cleaned with boiling water to keep the animals healthy, packed with swamp earth.[5] For long sea voyages enterprising merchants built mobile artificial environments inspired by the idea of a 'portable swamp'. Such containers consisted of earth or clay kneaded into dough, covered in a cloth with a minute hole for ventilation. Safely housed, leeches were shipped vast distances, from the Old to the New World or, as Old World sources dried up, from the New to the Old. By the 1870s leeches were regularly shipped from Australia to Britain, arriving 'in good condition' and pronounced 'very fine'.[6]

An ornate blue gilt earthenware leech jar designed for prominent display by apothecaries, communicating the prestige of the establishment to prospective customers.

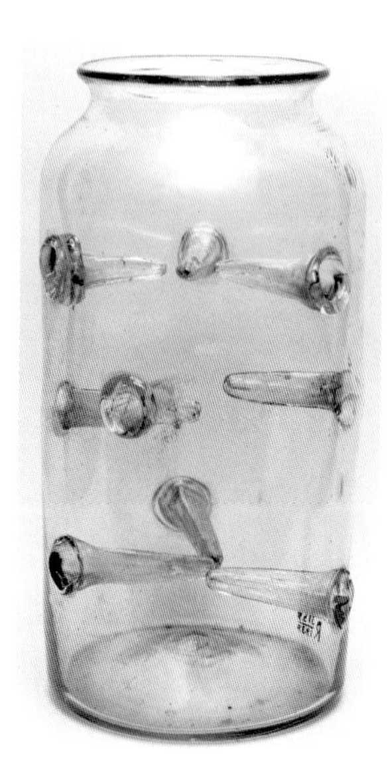

This novel glass leech jar incorporates a number of inward-pointing hollow tubes that terminate in tiny air holes, ensuring a supply of fresh air. Leeches could attach themselves to these structures, allowing for exercise and entertainment.

Wholesalers, too, built enormous artificial swamps to store leeches prior to sale. On arrival in their country of destination, leeches were first pooled within 'purging ponds', which allowed the animals to recover from their journeys and digest whatever remained of the meal they had been fed prior to departure. This was also designed to allow leeches to adjust to the new climate, a process that could require leeches to spend up to a year in the 'purging pond'. Given the time-consuming complexity and demanding labour involved in the leech trade, it is little surprise that those apothecaries maintaining a reliable supply of leeches were held in great esteem. As leeches grew in prestige, so too did their place within the apothecary. By the 1820s, leeches had come to indicate the overall quality of an apothecary to would-be clients. Early leech jars had been made of simple white pottery, sometimes lettered in black with 'leeches', and were of no real

Clay leech storage jar. There were many designs of leech jar; all featured distinct patterns of holes to allow leeches to breath and maintain their health. This example has a heavy iron clasp to prevent the leeches escaping.

An elegant leech jar designed for display within an apothecary.

note. By the height of the leech craze, however, they had come to be homed in ornate jars up to two feet in height, brightly coloured, highly decorated and prominently displayed on the pharmacist's shelf. A good leech, after all, required a good home. How embarrassing it would be to the apothecary who possessed such a jar but let it stand empty.

The growth of leech trade and transportation was not without its peculiarities and problems. In Britain, some apothecaries attempted to make use of the then newly established postal service, not without incident:

> Among the strange packages which came by the General Post to Stamford on Tuesday was one containing 200 leeches. On emptying the letter-bag the postmaster found to his dismay scores of the reptiles crawling about it: some

> had crept into letters, and others immediately seized his hands, and it was with great difficulty the dismayed official escaped a bleeding: they were all secured after a diligent search.[7]

The frequency of such accidents led the Post Office to ban the mailing of leeches. Escapees were also a problem in the early days of leech shipping. At British ports it was not uncommon to see merchant vessels arrive with decks swarming with leeches that had escaped their confinement during the journey.[8]

One consequence of the booming leech trade was that surgeons, doctors, merchants, patients and druggists began to differentiate medicinal leeches by their place of origin. National

'Awful Occurrence': a 19th-century engraving of the panic that ensues when leeches escape from their jar on an omnibus.

AWFUL OCCURRENCE.

Chorus of Unprotected Females. "CONDUCTOR! STOP! CONDUCTOR! OMNIBUS-MAN! HERE'S A GENTLEMAN HAD AN ACCIDENT AND BROKE A JAR OF LEECHES, AND THEY'RE ALL OVER THE OMNIBUS!"

A 19th-century cage for capturing leeches, probably used by leech wholesalers.

characteristics, many believed, indicated medical prowess, leading to some leeches being more commercially desirable than others. Traders who sourced leeches from southern Europe, for instance, claimed the climate gave their animals a superior growth and metabolism, making them more appropriate for medical use than those of temperate northern climes. Leeches were also anthropomorphized, having national character and behavioural traits projected upon them. In the 1840s, for example, British physicians feared that Greek sellers were feeding oxblood to their leeches to fatten them, in the belief that English patients preferred their leeches to resemble the fat and stocky image of John Bull.[9] National traits were freely read into leeches, shaping their use and care, as one Norwich-based leech trader explained:

the Prussian Leeches bite the best . . . the French leeches are slow biters, and . . . the Portugal leech is not fit for use till it has been kept in the ponds. These varieties attack and even destroy each other; the Hungarian, particularly, attacking and biting the Prussian when they are first put together.[10]

With British and French supplies almost exhausted, the Prussian and Hungarian leech had risen to dominate the European leech trade by the mid-nineteenth century, making such considerations increasingly important for leech wholesalers.

Traffic in leeches became a transatlantic trade as a result of European immigration to the United States. In America, European leeches were highly prized as the native species, *Hirudo decora*, was generally thought to possess a shallower bite, drawing comparatively little blood. Herman Witte, a recent German immigrant, was the first to regularly import European leeches,

16 & 17, ST. PETER STREET, HACKNEY ROAD, N.E.,

Mr John Green — LONDON, May 12 186

LEECHES. Wholesale and for Exportation.

LEECHES. Wholesale and for Exportation.

HAMBRO' SPECKLED.

OFFICINAL GREEN.

AQUARIA FOR LEECHES.

Bought of FITCH & NOTTINGHAM.

Post Office Orders to be made payable *Hackney Road* Office *only*. Cheques to be crossed London and County Bank.

57 Choice Leeches

(Leeches for Sutton)

Invoice from Fitch & Nottingham for 57 'choice' leeches sold to businessman Mr John Green, 12 May 1870.

establishing a business in New York City in 1839.[11] The growth of leeching in America again revealed the importance of being able to distinguish between species. Difference in appetite was already an important consideration when prescribing the correct number of leeches, and the dramatic discrepancy in appetite between European and American leeches took the inexperienced by surprise. In 1876 an outraged father, whose daughter was being treated with a mere twelve Swedish leeches, demanded 100 leeches be used on the basis of his own prior experience with American leeches. 'And had you applied one hundred of my leeches what would have been the result?' enquired the surgeon once the child had recovered. 'Death' was the shamed response of the father.[12] America's apparently inexhaustible demand for the European leech placed such strain on the global leech trade that some came to resent the U.S. for having exhausted the world's supply of this 'irreplaceable medical apparatus'.[13]

CAPITALIST LEECH

Western Christendom has long drawn on leeches to represent human greed, following Proverbs 30:15, which warned that the 'leech has two daughters. "Give! Give!" They cry.' The commodification of *H. medicinalis*, however, gave new life to the leech as metaphor for avarice. In the dawning age of capitalism, leeches appeared to embody wealth. Blood and capital became entwined as money was increasingly seen to be the 'blood' of capitalism. Just as a body is healthy when blood circulates vigorously about it, economic health requires the unimpeded circulation of money about the national economy. However, where the medicinal leech was an aid to bodily health, the economic leech was cast as a parasite upon capitalism. The caricaturist George Cruikshank used a leech motif in his illustration of 1811 depicting John Bull

The biblical reference to the greed of horse leeches inspired the zoomorphic transformation of two Irish patriots into leeches, who cry: 'Give! Give!' to support 'Poor Paddy and the children of the horse-leech'. *Frank Leslie's Illustrated Newspaper*, 4 March 1871.

being ruthlessly bled. In a German lithograph from 1848 leeching becomes a metaphor for the way in which the German princely houses fed off the people. The message conveyed is one of republican sentiment, implying that the German Confederation was being drained by Austria, a larger imperial power threatening the smaller Germanic states.

Critiques of capitalism from the nineteenth century onwards have regularly invoked the leech as a watchword for economic greed. In 1842 the *Penny Satirist* drew heavily on imagery of the medicinal leech in portraying the avaricious nature of capitalism:

> The leech has two ends . . . One of these ends is a mouth and the other is not a bottom. So that it keeps everything it gets, and hoards it all up in its bags, and when it is sucking the bags are filled with blood. They are just like little purses – and when seen with a microscope with the red blood in them they are just the very images of bags of gold. So that the inside of the leech is like a miser's hoard, and there is no outlet and no inlet, the creature keeps it all to itself, and distributes nothing . . . and nourishes itself without labour . . . It is a capitalist.[14]

Leeches were also mobilized to critique traditional forms of wealth. In 1857 *Reynold's Newspaper* criticized exuberant living, attacking aristocrats for draining the 'blood of hardworking millions'.[15]

The physician, signifying the German National Assembly of 1848, places one final leech on the long-suffering patient, who represents the German people. This final leech is Archduke John of Austria, who had been elected German Imperial Regent on 29 June 1848.

George Cruikshank, 'The Blessings of Paper Money; or, King a Bad Subject', 1811, coloured engraving. The two physicians represent parliamentarians responsible for a bill to depreciate the value of British bank notes, threatening to drain the wealth of the nation. Napoleon reveals his parasitical nature by withdrawing a pan filled with gold coins from underneath John Bull.

Similarly, a provincial British newspaper warned of a 'tribe of human leeches' that had transformed England into a country ruled by a 'government [of] phlebotomy'. With a distinctly republican sentiment, the author continued: 'there is the hereditary leech, the Court Leech, the Parliamentary leech, the legal leech, the clerical leech, the gaol leech, and scores of other families'.[16] Whilst radical politics drew on the perceived behaviour of leeches to critique the powers of the day, be they capitalist or aristocrat, the true victim was the reputation of the leech itself.

Comparable themes were woven into the figure of the vampire, an anthropomorphized leech who possesses a leech-like thirst for blood alongside a human appetite for money. In nineteenth-century literature leech, vampire and capitalist were woven together around the common theme of blood and money. Monstrosity, in early vampire stories, is conveyed not so much from bloodsucking but an insatiable desire for money, such as in James Malcolm Rymer's *Varney, the Vampyre; or, the Feast of Blood: A Romance*. Sir Francis Varney, an urbane gentleman who happens also to be a vampire fallen on hard times, preys upon

Dracula famously travels within a wooden casket filled with earth, mirroring the way leeches were packed for transport across the world in the 19th century. Still from Francis Ford Coppola's *Dracula* (1992).

the wealth of the English landed elite. His bite, comparable to a 'million and half leeches rolled into one', was used to drain victims not only of their blood but their capital.[17] Informing the story were concerns about capitalism and the emerging urban industrial order, which were threatening traditional aristocratic culture. Even Karl Marx, that famous critic of capitalism, exploited the metaphor:

> The prolongation of the working day beyond the limits of the natural day, into the night . . . quenches only in a slight degree the vampire thirst for the living blood of labour. To appropriate labour during all the 24 hours of the day is, therefore, the inherent tendency of all capitalist production.[18]

The association between blood and money, body and economy, leech and capitalist, has subsequently entered into the collective Western imagination. In 1979, for example, Freddie Mercury came to believe Queen's ex-manager had held back the full profits from the band's early albums. In response, Mercury penned a song which addressed someone who was sucking his blood like a leech, taking all the singer's money and still wanting more.[19]

TRADED LEECH

The leech, however, is not just a metaphor for capital. It *is* capital. Driven by the nineteenth-century 'leech mania', leeches became a global commodity. Quite literally, there was money in leeches. Hitherto undesired wetland, particularly in Marseilles, Algiers, Smyrna (now Izmir), Egypt and Trieste, suddenly became much sought after as, on the back of the medical blood rush, merchants sought their fortunes in swamps.

Early 1830s Hungary established itself as the pre-eminent supplier of European leeches, exporting up to 60 tons of leeches annually (roughly 8 to 12 million leeches). Such was the boom in Hungarian leeches that the average trader yielded 280,000–400,000 forints annually, an income equivalent to the richest aristocrats of Hungary and way beyond the ten to twelve forints of a peasant household.[20] Hardy leeches such as the Swedish, which obtained a reputation for surviving transatlantic journeys, became particularly sought after outside their homeland. Worth only two öre each in their native country, on the international market, driven by American demands, a single Swedish leech fetched up to 100 öre by 1860. The booming leech trade stretched beyond the West, with British and French colonial expansion creating new markets for leeches across the far-flung outposts of empire. Difficulty in competing with the powerful European and American demand for leeches led colonialists to experiment with locally sourced species. India, possessing an indigenous and ancient tradition of using leeches in the Ayurveda medical system, was perfectly placed to dominate the Southeast Asian

Farmed medicinal leeches caught in a sieve in preparation for supply to customer.

leech trade that arrived with Western colonial expansion. The Indian city of Pondicherry developed a robust leech-based economy, exporting local species, likely *Hirudinaria manillensis* and *Poecilobdella granulosa*, to outposts as distant as the British/French colony of Mauritius.[21]

Where there was money, crime soon followed. Incidents of leech piracy tended to follow the decline of local stocks. Norwegians, having fished their native leeches to extinction, raided the pools of Sweden, threatening to devastate Swedish stocks whilst imperilling relations between the two recently unified countries.[22] What was first welcomed as a boon to the economy soon became a threat to national health as overfishing for export threatened to eradicate national stocks. Moldavia, which joined the exportation boom in 1835, exhausted its indigenous leeches within

Leech-gathering was an important part of the Yorkshire rural economy. In this illustration of 1814, women are plucking leeches from their legs and placing them in water-filled firkins for transportation across England.

two years. Italian, Spanish and Portuguese stocks were similarly exhausted by the mid-nineteenth century. As the health of citizens was increasingly recognized to depend on access to leeches, diminishing national stocks became an ever more pressing state concern, leading governments to first restrict and then ban exportation. In 1823, for instance, the Kingdom of Hanover outlawed the export of leeches to preserve domestic populations. Austria, by contrast, managed exportation by issuing exclusive five-year rights to trade in leeches to two Viennese dealers in 1827. Swedish medical men unsuccessfully petitioned the state to impose an export ban in 1831. In 1848 Russia imposed high tariffs on the exportation of leeches in conjunction with an annual ban on leech gathering between the months of May and July, but all this achieved was a boom in leech smuggling, with

German dealers flagrantly building large leech ponds just across the border to store contraband animals.[23]

In parts of Europe such as Britain and France, where native leeches had been all but eradicated, dependence on the importation of leeches became an area of urgent state concern. Both countries experimented with taxing the importation of leeches, which established a new and lucrative revenue stream. However, the intended aim was to reverse the trade deficit by encouraging the domestic production of leeches, particularly important in times of war.[24] So began the farming of leeches on an industrial scale.

FARMED LEECH

By the 1830s the world was on the precipice of a leech crisis. In many countries the poor could no longer afford to use leeches, whilst hospitals were struggling with the rising costs of leeching. Recognizing an opportunity in the growing economic and medical value of leeches on the one hand, and their scarcity in the wild on the other, entrepreneurs sought ways of artificially producing leeches. Their efforts were encouraged by state sponsorship, such as the reward of $500 offered by the United States government in 1835 to anybody who succeeded in breeding the European leech in America. In 1834, when disease devastated native Swedish leech stocks at a time when they were sorely needed to combat a cholera outbreak threatening the population, the government responded by funding the Karolinska Society to establish an experimental leech farm outside Stockholm. Despite having enrolled the leading scientific minds of the time, this endeavour was a failure because not enough was yet known about the natural habits and needs of leeches to allow them to be successfully bred in captivity.[25]

Leech House, Bedale, North Yorkshire. This unique 19th-century structure was built to store medicinal leeches for supply to apothecaries.

The art of 'hirudoculture', or farming leeches, was first mastered by the French following the establishment of a government committee to investigate methods of artificially propagating leeches. Furthermore, the state, in conjunction with the Académie Nationale de Médecine, established a prize fund for anybody who successfully demonstrated a working 'leechery'.[26] Temporarily storing leeches for medical use was quite different to maintaining permanent healthy populations that would breed in captivity. It took many years to develop successful propagation techniques because so little was known about the reproductive behaviour of wild leeches. A further problem was the difficulty of feeding the large numbers required for the mass production of leeches. It was a farmer from the Gironde region of western France, M. Béchade, who solved this problem in 1835 by cutting open the bodies of horses, cows or donkeys and driving them into his marshland farm. This technique was an ideal means to feed leeches and, if the animal was walked through and out of the marsh, it also served as a method of harvesting.[27]

Despite consuming 700 to 750 horses a year, Béchade could make a healthy profit and was one of the first to demonstrate the power of hirudoculture. Revealing a new fertility in what had been poor swampy farmland, leech propagation created a bountiful new agriculture in wetlands rich with leeches.[28] Béchade's leech farm was so successful it remains in business today, trading as Ricarimpex Sangsues Médicinales.

Hirudoculture was celebrated as a modern and technical response to the depletion of leeches in the wild. With leech farms appearing across the country in ever more elegant designs, France became renowned for the most sophisticated techniques of artificial leech propagation, to the extent that farms became integral parts of French urban reconstruction. Of course, leecheries did not spring fully formed into the suburbs of French cities: there were many problems to overcome. Until efforts to maintain consistent artificial climates were established, leech stocks could be decimated by seasonal change. Inadequate ventilation combined

A valley in the region of Belfort, France, after being transformed into a modern leech farm in the late 1850s.

with overcrowding driven by the desire to maximize production also led to the prominence of leech diseases.[29] Vermin, particularly rats and birds, were also a problem. In 1854, 200,000 leeches were lost from one farm when a flock of wild geese stopped for a brief rest in the inviting new pools that had appeared on their usual migration route. How many of these leeches were consumed by the geese, and how many absconded by attaching themselves to the bellies of the birds, is difficult to say. Poaching also caused setbacks. To deal with these mounting issues, one hirudoculturist,

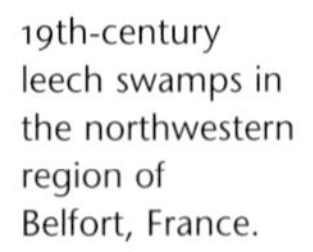

19th-century leech swamps in the northwestern region of Belfort, France.

Landscaped breeding pools created by E. Devès in the French region of Ambe, 1855.

Hirudoculturalist at a French leechery packing medicinal leeches in clay for transport, 1905.

M. Borne, built a large lighthouse at the centre of his farm in which he employed armed men to stand guard day and night. Innovation in hirudoculture was driven by contested debates over the best ways to adapt nature to serve the needs of medicinal commerce without corrupting the health or utility of leeches. Borne possessed a protective sentiment toward his leeches, treating them with the utmost 'care and tenderness'. He built an artificial incubator for their cocoons and developed a special diet of cow's blood to help the young leeches grow, blurring boundaries between the natural and the artificial by adopting a paternalistic role towards what he affectionately called his 'pupils'.[30]

Other nations proved keen to emulate French hirudoculture. One of the largest leecheries outside France was in Australia, constructed by R. P. Negus, a businessman-cum-'Leech Conservator', with considerable support from the government of Victoria. Within a short time Negus transformed twelve acres of swamp by the Murray River into a bountiful leech farm, breeding hundreds

A French leechery designed around a spiral pool, *c.* 1855.

of thousands of 'the beautiful Murray leech' for supply to the American market.[31] The transformation of the Australian wilderness into a site of commercial production appealed to the American collective imagination, in which the frontier still loomed large, with the American press announcing that the Australian method would supersede that of Europe by providing an 'unlimited' supply.[32] Curiously, the British were less impressed. Despite the Australian leech trade having reached 'very extensive dimensions', being able to supply 300,000 to 500,000 leeches a month, the *British Medical Journal* snootily commented that 'they are intended for the London market: but not, we hope, to be used in London'.[33]

Overall, the British were not keen on artificial propagation. Writing of his visit to the French leecheries near Bordeaux, the best one British critic could say was that 'hirudiculture is a tolerably disgusting business'.[34] Disgust, however, was less focused on the leech than on the means of feeding them. The use of cut horses to feed leeches was an unpleasant sight to witness: loss of blood reduced the animals to an exhausted state, often causing

'Driving Horses out of a Leech Swamp, near Bordeaux, France', 1866, engraving.

them to collapse, forcing farmers to brave the swamps to haul out the bodies once they had been bled dry. To overcome this problem the French developed a purpose-built box within which horses were strapped and kept standing whilst their blood was consumed. Hirudoculturists were keen to preserve horses for as long as possible due to the cost of replacement. There is little doubt these animals suffered terribly, as having been introduced to their new work horses would frequently refuse to re-enter pools until physically forced to do so.

The plights of these animals moved early animal advocates to bemoan the 'horrors' and 'agonies' inflicted upon horses within the artificial swamps created by hirudoculture. If medicinal leeches were fed on 'agonizing and often diseased horses', imprisoned in compartments where leeches fed on them 'instantly by thousands', one outraged reporter demanded how we could know that the leeches themselves would not spread infection and agony when used in medicine.[35] Growing Anglo-American sentiment towards animals inspired regular outbursts against the methods of artificially propagating leeches. 'It may be that the interests of the healing art require such appliances', concluded one such piece in 1866, 'but one could hardly look upon a leech without abhorrence if aware that it had been grown by this process.'[36]

Moral outrage against artificial leech production rehearsed wider public revulsion over the mistreatment of animals for the benefit of human health. Sensational media reports warning of man-made pools filled with thousands of predatory leeches can be read against the better-known protests against vivisection. A vivid example told of a boy employed to drive horses into leech ponds who, while waiting for the leeches to feed, fell asleep on the job. The unnaturally corrupted leeches, it was reported, saw an opportunity in the sleeping body so close to their shallow pond,

In Istanbul leeches for medicinal use are commonly sold directly to the public in markets. In 2011 leeches sold at 1 Turkish lira each (about 40p).

fastening upon him in their hundreds and bleeding him to death.[37] Similar stories were circulated through Anglo-American newspapers from the mid- to late nineteenth century, illustrating the ambiguous revulsion many held for the industrial manufacture of leeches. The abhorrence towards leech breeding was related to the use of live animals, mostly horses but also donkeys which, in an age where violence to animals was increasingly viewed as barbaric, could hardly be tolerated. Affection towards the leech, however, was notably absent. The suffering of the leech, removed from nature, confined and bred in overcrowded artificial environments and forced to feed on old, diseased animals until needed by human medicine, received little sympathy. In these tales, the leech was a largely unacknowledged victim.

Yet it was the human need for medicinal leeches that had led to these animals being all but eradicated from their natural homes. With their craving for profit unsatisfied, enterprising capitalists imprisoned and bred leeches for their own gain. But leeches were not just physically enslaved, they were also conceptually imprisoned. The avaricious desires driving their entrapment into capitalism, together with all the worst characteristics of human nature, were, through metaphor, transferred into the figurative body of the leech. In this way leeches were denied victimhood, instead coming to symbolize the very worst characteristics of capitalism. Leeches, ironically, came to represent that which had exploited and opressed them. As Marx noted, appropriation is the inherent tendency of capitalists. Perhaps, however, it is time to rethink and release leeches from this association which, arguably, they so little deserve.

4 Mechanical Leech

There have always been those who prefer the predictable to the unpredictable, who dream of substituting life with ingenious mechanisms of their own making. Accordingly, wherever leeches have appeared in human culture, there have been those who have tried to better nature by building a superior, mechanical leech. For every attempt to replace leeches, however, there have been corresponding efforts to enhance their natural abilities by crafting better ways of working with them. Here, rather than being obstructive, the characteristics of life (such as variation, empathy and unpredictability) are harnessed as virtues necessary for creative cross-species collaboration. In this chapter we explore both these approaches to incorporating leeches into mechanical worlds.

METEOROLOGICAL LEECHES

Many nineteenth-century households kept leeches in anticipation of medical need. Housing leeches domestically required only the most basic of mechanisms: a jar, some soil and regular infusions of fresh water. Whilst waiting for medical emergencies, however, leeches were expected to perform other household duties. First and foremost was weather forecasting. Edward Jenner (1749–1823), in a much-reproduced poem, recounts numerous 'Signs of Rain', one of which was the movement of leeches:

> The leech. Disturb'd, is newly risen
> Quite to the summit of his prison . . .
> 'Twill surely rain – I see with sorrow,
> Our Jaunt must be put off to-morrow.[1]

Jenner knew well the abilities of leeches to sense changes in the weather. Like many British physicians of his day, he enjoyed studying nature and was well versed in natural history. His study of the parasitical nesting habits of cuckoos, for instance, won him scientific credibility and election to the prestigious Royal Society in 1788.[2] As a physician, Jenner was a connoisseur of leeches, too, but unlike the cuckoo, he did not write of them as parasites.

Today Jenner is known as a pioneer of vaccination. By demonstrating that inoculation with cowpox provided protection from smallpox, he laid the foundations for the modern science of immunology and the eradication of this horrendously disfiguring disease in the twentieth century. In his day, however, the idea that cows and their infections could play a role in medicine was highly controversial. Jenner's enrolment of the cow into medicine was initially met with revulsion. Popular fears of this new and foreign relationship between human and cow bodies contrasted sharply with the established cultural acceptance of the medicinal leech. Yet leeches were not only medical companions, but served as a means to read the weather.

Keeping leeches for use in phlebotomy, it is little surprise Jenner noticed that when rain approached his leeches disturbed themselves from the soil at the base of their jars and swam to the top of their 'prison'. The idea that leeches should be gifted with the ability to sense changes in the weather made perfect sense within nineteenth-century culture. Mariners, shepherds and farmers, whose lives were intimately bound up with their meteorological environment, had long used animal behaviour as a way to read and predict the weather. Before meteorology became a modern science, weather prediction was considered to be a matter of instinct, not reasoning; thus animals, whose instincts were so much more attuned to nature than man's own, were considered to be gifted weather prognosticators.[3] Few were as talented as the leech.

In 1806 the popular *Monthly Magazine or British Register* recommended leeches as meteorological companions. One writer explained:

> A phial of water, containing a leech-worm, I kept on the frame of my lower sash-window, so that when I looked in the morning I could know what would be the weather of the following day.

James Gillray, 'The Cow-pock; or, The Wonderful Effects of the New Inoculation!', portraying contemporary fears of the destabilization of human and animal identities through Edward Jenner's new vaccine-based medicine, 1802, etching.

> 1st. If the weather proves serene and beautiful the leech lies motionless at the bottom of the glass and rolled together in a spiral form.
> 2nd. If it rains either before or after noon, it is found crept up to the top of its lodging and there it remains until the weather is settled.
> 3rd. If we are to have wind, the poor prisoner gallops through its limpid habitation with amazing swiftness and seldom rests till it begins to blow hard.
> 4th. If a remarkable storm of thunder and rain is to succeed, for some days before it lodges almost continually without the water, and discovers uncommon uneasiness, in violent throws, and convulsive like motions.
> 5th. In the frost, as in clear summer-weather, it lies constantly at the bottom. And in snow, as in rainy weather, it pitches its dwelling upon the very mouth of the phial.
> What reason may be assigned for them, I must leave to philosophers to determine: the one thing is evident to everybody . . . that the change of weather, even days before, makes a visible alteration on its manner of living.[4]

William Cowper (1731–1800), the English poet and writer of hymns, was one who took up this practice. Having obtained a leech for sixpence, then 'a groat over the market price', Cowper nonetheless believed his leech to be 'an invaluable acquisition' as 'no change of weather surprises him, and that in point of the earliest and most accurate intelligence, he is worth all the barometers in the world'.[5]

Leeches' ability to sense changes in the weather was understood to be directly related to their unique uses in medicine. For centuries the leech had served an integral position within humoral theory, a system of medicine that placed great emphasis upon the relationship of body to environment. Any sudden change, whether internal to the body or external in the wider climate, was believed to unbalance the humours, leading to sickness. Within this medical world the leech held the power to intervene between body and climate, rebalancing the humours and returning the sick to health. That abrupt changes in the weather could cause disease was well established, as in humoral medicine each humour corresponded with a season (blood with spring, yellow bile with summer, black bile with autumn and phlegm with winter). Although in the nineteenth century humoral explanations declined, disease continued to be generally linked to the environment through related ideas of 'miasmas' and 'bad airs'. Against this context, in which body and space, climate and health, were deeply interconnected, and where the leech held the power to manage these relations, it was comparatively easy to believe that these animals possessed the ability to read and predict the weather.

THE TEMPEST PROGNOSTICATOR

Fifty years before Bram Stoker associated Whitby with Count Dracula, the small coastal town was popularly known as the home of the 'tempest prognosticator', a fantastically named device for the accurate forecasting of the weather. Inspired by Jenner's poetry, Dr George Merryweather, physician and resident of the small North Yorkshire fishing town, had built the apparatus to enhance the ability of leeches to forecast the weather. Merryweather (an appropriate name if ever there was one) described his machine as 'an atmospheric electromagnetic telegraph conducted by animal instinct', capable of predicting future storms 'with unerring certainty'. [6]

Twelve leeches formed the beating heart of the tempest prognosticator. Each was housed in an ornate jar incorporating a

Design for the Tempest Prognosticator.

Dr George Merryweather's Tempest Prognosticator on display at The Whitby Literary and Philosophical Society Museum.

whalebone lever that would be triggered when a leech ventured to the top of their home. The lever was connected to a central dome, which in turn would be connected to a telegraph. The tempest prognosticator served as a fully automated instrument, enabling a single leech to anticipate a change in climate and to communicate it across any distance. Within such a machine, Merryweather boasted, 'a little leech, governed by its instinct' could 'ring Saint Paul's great bell in London as a signal for an approaching storm'.[7] The integration of animal instinctual sensibilities with human technological innovation seemed reasonable to educated Victorians, who believed in both the infallible precision of instinctive sensations and the progress of civilization through science and technology. Leech and telegraph could be united because mechanical metaphors were increasingly dominant within physiological understandings of how bodies worked. Thus the leech's nervous system could be 'wired' into the telegraph, the product of nature integrated with human artifice.

But within this machine, can we properly talk of leeches being meteorological companions? Had leeches become machines? Or were they cyborgs, born long before their day? Merryweather, at least, believed the leeches to be his living companions:

> If I were to say that leeches were capable of attachment, you might well say that everybody knew that fact; I therefore am obliged to say that leeches are capable of affection, for after they become acquainted with me, they never attempt to bite me; but some of them have, over and over again, thrown themselves into graceful undulations when I have approached them; I suppose as an expression of their being glad to see me.[8]

This affectionate relationship was mutual. Merryweather came to care deeply for the leeches, whom he thought of as his 'comrades', and did his best to provide for their well-being. He reasoned, for instance, that leeches could live in isolation without undue suffering because, being hermaphrodites, they did not desire a sexual partner. However, Merryweather was still keen to provide for their emotional and social needs. To prevent loneliness, he adopted a circular design to allow each jar to be positioned so 'that the leeches might see one another, and not endure the affliction of solitary confinement'.[9] In this way, the materiality of the tempest prognosticator was shaped to provide a comfortable home for Merryweather's fellow workers.

Working closely with leeches taught Merryweather that each was an individual. He had initially assumed that 'leeches kept separately, and placed under similar circumstances, would simultaneously give indications of thunder'. Merryweather quickly discovered 'this not to be the case: some appeared to be more sensitive and more prophetic than others; and some appeared to be absolutely stupid'. By rendering leech behaviour comparable, the tempest prognosticator provided leeches with a means to express their individuality, as some demonstrated their utility and enthusiasm for the work by accurately prophesying all changes in the weather whilst others responded only to some weather conditions and not others. A variety of leeches of different skills were required to overcome individual preferences, leading Merryweather to choose twelve because this number connoted wisdom (derived from Jesus' choosing of twelve disciples and exemplified in the English practice of appointing twelve jurors in a criminal trial). It was not for nothing that Merryweather collectively referred to his leeches as a 'jury of philosophical counsellors'.

The Great Exhibition of 1851 brought the tempest prognosticator to public prominence when it was displayed as an example of the superiority of British power, technological ingenuity and industrial strength. But did the tempest prognosticator work?

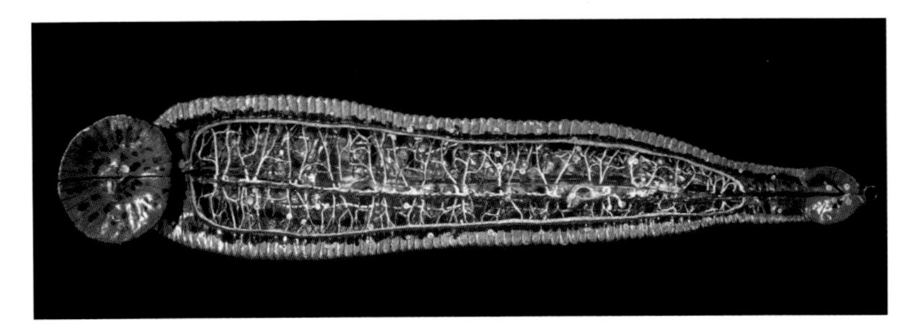

A papier-mâché anatomical model of the internal organs of the medicinal leech created by Louis Auzoux, *c.* 1850.

To prove it did, Merryweather made innovative use of another recently developed mode of communication: the Penny Post. Whenever his leeches predicted a storm in 1850, Merryweather had despatched a forewarning letter to Henry Belcher, then president of the Whitby Philosophical Society. Using the Post Office, which date- and time-stamped the letter, Merryweather collected incontrovertible evidence of the leeches' prognosticating abilities.[10] With naval power and maritime supremacy increasingly central to the security of the British state and empire, not to mention global trade, the ability to accurately forecast the weather became a matter of national importance.[11] Did the leech have a role to play in protecting British imperial aspirations and economic dominance? Some thought so. Reporting the hubbub about the machine, one newspaper exclaimed that it might save

> an immense annual loss of life and property and would be worthy of that prominent and easily accessible place at the Great Exhibition.[12]

Lloyd's of London, the leading insurance company of the day, with an obvious interest in preventing the loss of shipping, agreed, running its own tests confirming the tempest prognosticator's accuracy.

Greatly encouraged, Merryweather designed five further versions of his instrument, each intended to be affordable and adaptable for different purposes. Hoping the tempest prognosticator would be widely used on ships, Merryweather petitioned the government to install prognosticators around the British coastline, creating a weather forecasting network. However, there were several rivals to the tempest prognosticator also on show at the Great Exhibition. Robert FitzRoy, famous for captaining the HMS *Beagle* upon which Charles Darwin voyaged, had invented his own weather prognosticator. His storm glass was a sealed cylinder containing a liquid that changed in appearance in line with changing climates. FitzRoy's storm glass was a simple instrument requiring little to no maintenance; in contrast, the tempest prognosticator may have looked majestic but it was also labour intensive. The leeches required feeding every few months and the regular changing of their water. When FitzRoy was appointed Meteorological Statist at the Board of Trade in 1854, the fate of the meteorological leech was sealed – the tempest prognosticator turned out to be their opulent final home. Although leeches lost their forecasting role in the late nineteenth century, their work has not been forgotten, inspiring, for example, Isobel Dixon's poem 'The Tempest Prognosticator' (2011).[13] Leeches as 'weathermen' live on today in our cultural memory, inspiring twenty-first-century minds to think differently about the world they live in.

THE MECHANICAL LEECH

Like any tradable commodity subject to the combination of restricted supply, increasing demand and governmental taxation, the cost of medicinal leeches grew exponentially, particularly in times of scarcity. This created a tension between the will to work with the natural rhythms of the leech and the need to appropriate the leech to serve the needs of medicine. Though some physician-surgeons took great care of their leeches, allowing them to naturally digest their food, this became increasingly difficult as demand for their use made it challenging to accommodate the leeches' notoriously slow digestive system. Few busy cosmopolitan surgeons could afford to use a leech only once or twice a year. Consequently it became common practice to force leeches to regurgitate their meals by squeezing them between thumb and finger, a practice that risked fatally damaging the animal. Another technique for making efficient use of leeches involved snipping their tails, causing consumed blood to flow out of the body. Rather than prescribing a large number of animals, one surgically altered leech could thereby do the work of many. Once enough blood was drawn, the leech was removed from the patient and, by holding together the surgical incision in the leech's body for a few moments, a healing process could be instigated to close the wound. In this way a single leech could be cut, set to work and reused many times.[14] This allowed the leech to be transformed into a bloodsucking machine, never becoming satiated. However, some saw this as a temporary solution to the scarcity and high cost of leeches. What was really required was a machine to replace the leech.

Inventors of 'artificial leeches' were keen to recast living leeches as 'repulsive and disgusting', claiming that 'delicate and sensitive persons find it difficult to overcome their repugnance to contact with the cold and slimy reptile'.[15] Patients who refused leeching because they had previously suffered a leech wandering

Jean-Baptiste Sarlandière's artifical leech, *c.* 1817–19.

within their body were also mobilized as a justification for the adoption of mechanical leeches. In such marketing, the living nature of the leech was increasingly reclassified from asset to problem. Whereas living leeches drew variable amounts of blood, with even the same individual drinking different rates on different occasions, mechanical leeches, it was claimed, would always perform uniformly. Similarly, where artificial leeches could be trusted to operate alone, living leeches had to be carefully managed, even refusing to obey their instructor at times! The idea of the mechanical leech was particularly attractive to hospitals, where annual bills for living leeches were sucking up budgets at an alarming rate. Vast savings were promised by replacing leeches with reliable, economical machines. In 1819, for instance, Jean-Baptiste Sarlandière, a French physician and inventor, calculated that his mechanical leech would free up the 1.5 million francs then being spent on live leeches across France each year.[16] Sarlandière, of course, had a product to sell; nevertheless, claims such as these contributed to a growing sense that the leech may not, after all, have a role to play in modern medicine.

Artificial leeches operated by using a 'scarificator' to cut open the skin before drawing blood through mechanical suction. Sarlandière's artificial leech, consisting of a glass ball with two tubes, one attached to the patient and the other at an angle that could be attached to a pump, generated suction by creating a vacuum that would draw out blood. Whereas Sarlandière's synthetic leech was intended to be a generic tool for the treatment of all diseases, Charles Louis Heurteloup designed a specialist mechanical leech for the treatment of headaches. Much like a conventional leech bite, Heurteloup's device made a small circular incision on the head, allowing precisely one ounce of blood to be drawn via a suction pump.[17] Heurteloup's leeches were amongst the most successful, finding a ready audience in the U.S.,

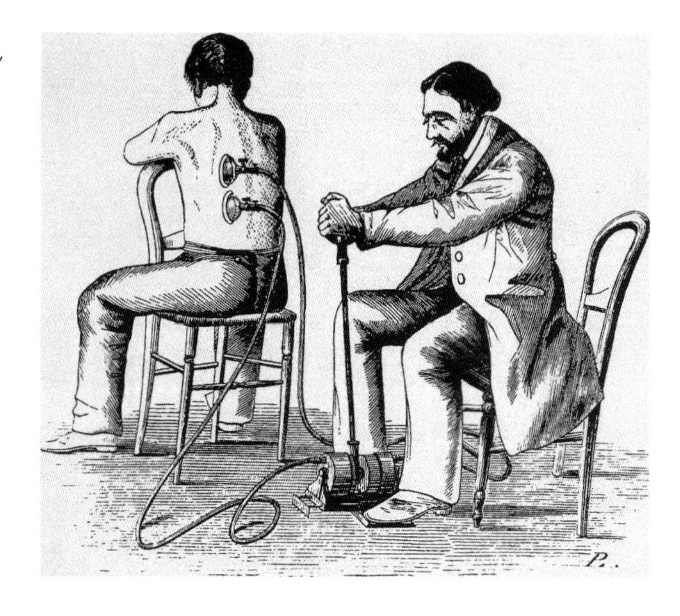

Damoiseau's artifical 'terabdella' ('large leech'), invented in *c.* 1862.

Artificial Leech produced by Tiemann, *c.* 1876, containing a blade that could puncture the patient's skin and a plunger that when withdrawn created suction to draw blood.

where they sold for $15 each. An advantage to mechanical leeches was that a design could be patented. In 1870, F. A. Stohlmann and A. H. Smith of New York patented their 'Artificial Leech' which consisted of a glass tube, either straight or curved so that the 'mouth' would appear on the side, a design thought to emulate the way a leech would 'hang'. This device, the patentees claimed, was not only cheap but any number might be applied at once and left to work of their own accord.[18] Mechanical leeches also featured at the Great Exhibition of 1851. Though drawing much less public interest than the tempest prognosticator, the Kidston Mechanical Leech was said to be the medical tool of the future.[19] W. Kidston & Co. claimed their mechanical leech was simple to use, capable of drawing any quantity of blood and well suited for use on ships or in any other location where procuring live leeches might be difficult.

Yet surgeons and physicians remained unconvinced. The mechanical leech might be a useful back-up if living leeches were unavailable but, other than that, there was little sustained enthusiasm once the excitement of a 'new' technological innovation had passed. 'The ingenuity of man', one experienced phlebotomist explained in 1876:

> can never furnish surgery with so perfect a bleeding apparatus as that wherewith nature has furnished the leech . . . You can imagine nothing more perfect. The leech itself is the most perfect phlebotomical apparatus for local bleeding conceivable . . . But the artificial leech . . . I keep these instruments because some prefer them, but as ingenious as they are, they can never be made, by any improvements, capable of supplanting the place of the leech in medicine.[20]

Many mechanical leeches were built, but are now found languishing in medical museums across the world. In the twenty-first century

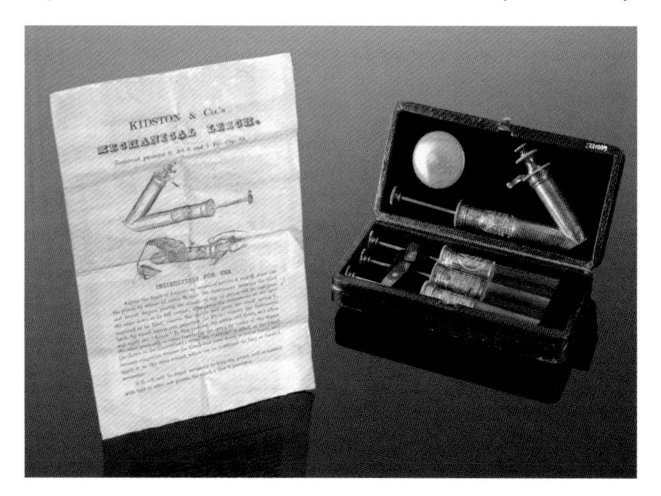

William Kidston and Co.'s Mechanical Leech, displayed at the Great Exhibition of 1851.

the quest to replace living leeches with artful human substitutes continues. Human inventiveness cannot as yet rival the sanguivorous expertise of the leech. So, might it not be better to think like Merryweather, accepting that many animals have abilities superior to our own? To work *with* nature rather than to subjugate, dominate and replace it with crude imitation? With a little humility, we might benefit greatly from those, such as leeches, who will lend us their skill, labour on our behalf, and ask so little in return. Indeed, we might even learn gratitude.

5 Wild Leech

In William Dalton's novel *Lost in Ceylon* (1861), leeches represent the savagery of an otherwise beautiful foreign landscape:

> How charming, how delightful, to spend one's life in such a place, says the reader. True, it would, if only there were . . . no land leeches, which, despite all effort, cling to your legs, and bleed you nigh to the weakness of death.[1]

By devouring travel literature and accounts of far-flung colonial outposts, nineteenth-century Western publics came to understand their societies through contrast with 'savage' cultures. Primitive tribes, inhabiting a wilderness populated by ferocious animals, came to symbolize the uncivilized world. Here, wild leeches took on the new role of cultural marker, embodying the limits of Western civilization. That erstwhile cosmopolitan companion *Hirudo medicinalis* was revealed to have much darker kin, and none more so than its bloodthirsty cousin *H. Ceylanica* (now *Haemadipsa zeylanica*): the 'land leech of Ceylon' (Sri Lanka).

In 1854 Joseph Dalton Hooker, botanist, famed explorer and close friend of Charles Darwin, described his encounters with leeches in the Himalayas:

> Leeches swarmed in incredible profusion in the streams and damp grass, and among the bushes: they got into my hair, hung on my eyelids, and crawled up my legs and down my back. I repeatedly took upwards of a hundred from my legs, where the small ones used to collect on the instep: the sores which they produced were not healed for five months afterwards, and I retain the scars to the present day.[2]

The only positive word Hooker had to say about wild leeches was the respite he gained in their absence. 'It is a well known fact', he explained, 'that these creatures have lived for days in the fauces, nares, and stomachs of the human subject, causing dreadful suffering and death.'[3] Such accounts were found throughout nineteenth-century travel literature. Ernst Haeckel, fond of representing his scientific expeditions as perilous adventures, reported his encounters with 'swarms' of the barbarous cousins of the European medical leech:

> I suddenly felt a sharp nip in my leg, and on baring it discovered a few small leeches which had attached themselves firmly to the calf, and saw at the same time half a dozen more of the nimble little wretches mounting my boot with surprising rapidity, like so many caterpillars. This was my first acquaintance with the much-to-be-execrated land-leeches of Ceylon, one of the intolerable curses of this

Tiger leeches (*Haemadipsa picta*) waiting for a passing meal.

> beautiful island, of all its plagues the worst, as I was afterwards to learn by much suffering. This species of leech (*Hirudo Ceylanica*) is one of the smallest of its family, but at the same time the most unpleasant.[4]

Exotic land leeches came at such speed, and in such numbers, they were impossible to avoid. They wriggled up the legs and through protective clothes with ease, even, some claimed, jumping some distance to reach their victims.

This was not the first time tales of wild leeches had entertained Western imaginations. Ibn Battuta, the fourteenth-century Moroccan explorer, described 'fierce leeches' lurking in the damp grasses of Sri Lanka. He also told of the traveller who, bravely ignoring the bites of leeches, was drained of so much blood that he died.[5] Much worse yet were the 'flying leeches' dropping from the sky onto those who ventured into the jungles. Today we have no knowledge of flying leeches; possibly Ibn Battuta had been misled by the land creatures' habit of crawling

Ernst Haeckel in Ceylon (now Sri Lanka) wearing 'leech gaiter' boots as a protection from land leeches.

up trees in order to drop on passers-by whose attention was absorbed by the siege from below. The deadly consequences of leech infestation proved a major problem to the British Army in its battles to bring Ceylon into the empire in 1815. Henry Allan, Deputy Inspector General of the British Army Hospitals, wrote of the many men lost to 'incessant attacks of the swarms of leeches', describing one case where a soldier was found to have 80 leeches feeding upon his body at once.[6]

When Sir James Emerson Tennent, Colonial Secretary of 'British' Ceylon between 1845 and 1850, published his study of

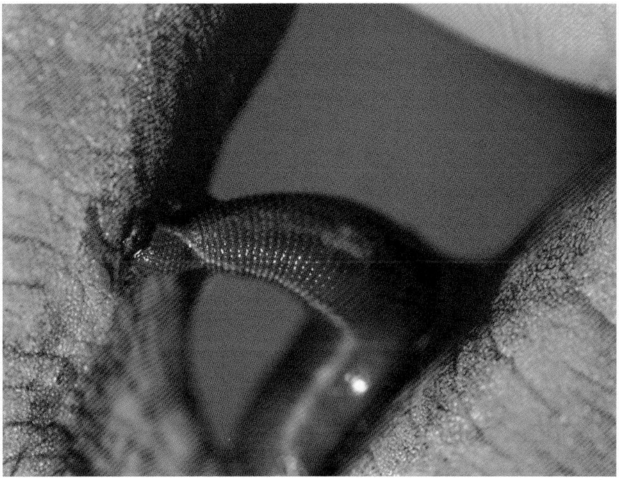

Leech feeding between two human toes.

the cultural and natural history of the country, he was the first to give a positive description of its indigenous land leech. He marvelled at how the little leech could become 'parched to hardness' during the prolonged drought of Ceylon's dry season, yet sprung back to life within minutes of the first rainfall.[7] They were also aesthetically pleasing, Emerson Tennent believed, possessing clear brown skin with two yellow stripes and a greenish dorsal one. Even the teeth were admirable, perfectly adapted for their purpose and 'quite beautiful'. Nonetheless, the land leech remained 'of all the plagues which beset the traveller in the rising grounds of Ceylon the most detested'.[8]

In contrast to other indigenous populations, human and animal, land leeches taunted British colonial aspirations, refusing to submit to its dominance. The pioneering nineteenth-century travel writer Constance Frederica Gordon-Cumming observed:

> the land leeches, which swarm in damp and luxuriant grass, have no tendency to fly from man. On the contrary,

This wooden mask was used in the Kolam plays of Sri Lanka. It represents Hevaya, a figure covered in bites and leeches, who was invoked within ancient fertility rituals and as a prelude to the exorcism of demons, which were believed to be the cause of disease.

the footfall of man or beast is as a welcome dinner bell, at sound of which the hungry little creatures hurry from all sides; and as each is furnished with five pairs of eyes, they can keep a sharp look out for their prey, which they do by resting on the tip of the tail, and raising themselves perpendicularly to look around. Then, arching their body head foremost, and bringing up the tail, they rapidly make for their victim.[9]

Being only two to three centimetres in length, and as thin as a knitting needle when unfed, land leeches can move at incredible speed. They use their muscular elasticity to effortlessly propel themselves through even the toughest of protective clothing, worming their way to the choicest parts of their host. Like all leeches they prefer the softest of skin, particularly the tender human neck, groin and genitals. Insignificant in size and skilful in their bite, victims rarely notice the arrival of their dinner guest until the leech has gorged to a size where the 'chill feeling' of skin upon skin becomes perceptible. In 1883 Haeckel recalled becoming suddenly aware, in the midst of enjoying a party, of an uninvited companion, having noticed a streak of blood running down his white evening trousers![10]

Land leeches are not the only members of their species to reside in Sri Lanka. *H. sanguisorba*, a leech resembling *H. medicinalis* but with twice the appetite of its European relative, was widely

Leech moving on human skin.

used as a medicinal leech in nineteenth-century Sri Lanka, harvested from a spring near the village of Tanyuttu in the Northern Province.[11] Whilst land leeches cannot claim to be the friendliest of the family to inhabit the island, nor are they the fiercest. This dubious honour belongs to the 'cattle leech', *Hirudinaria granulosa* (formerly *Haemopis paludum*), a species of leech that has lost the capacity to bite through the outer skin of mammals. Instead, these leeches feed on animals that wade into rivers and pools to drink, allowing themselves to be swallowed, or swiftly swimming up the nasal cavities of unsuspecting hosts. Once inside they find a comfortable lodging in throat or nostril, attaching to feed from the much softer internal flesh. So many might take up residence in a single animal that, having gorged upon blood, their expansion inadvertently suffocates their host. Not that this appears to be a concern of the cattle leech, which, as Emerson Tennent warned, is 'so tenacious . . . that even after death they retain their hold for some hours'.[12] Lacking the knowledge of indigenous people, many Europeans suffered a nasty encounter

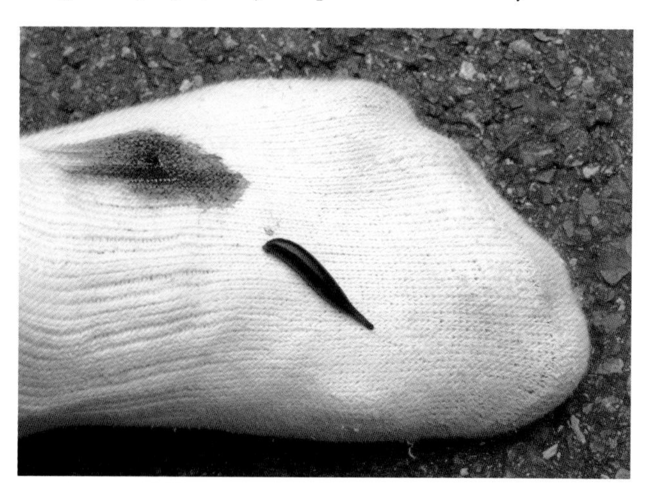

A Thai land leech (*Haemadipsa zeylanica*) navigating a tourist's foot.

with *Hirudinaria granulosa* when drinking cold water from an inviting pool on a hot Sri Lankan day.

However, not all who encountered the wilder cousins of the European leech were repelled. In his memoir Robert Knox, a seventeenth-century sea captain in the British East India Company who spent nineteen years imprisoned in Ceylon, told how:

> Leaches [*sic*] seize upon the Legs of Travellers; who going barefoot according to the custom of that Land, have them hanging upon their Legs in multitudes, which suck their blood till their bellies are full, and then drop off. They come in such quantities, that the people cannot pull them off so fast as they crawl on. The blood runs pouring down their Legs all the way they go . . . Some therefore will tie a piece of Lemon and Salt in a rag and fasten it unto a stick, and ever and anon strike it upon their Legs to make the leeches drop off: others will scrape them off with a reed cut flat . . . But this is so troublesome, and they come on again so fast and so numerous, that it is not worth their while and generally they suffer them to bite and remain on their legs during the journey and they do the more patiently permit them because it is so wholesome for them. When they come to their Journeys end they rub, all their Legs with ashes, and so clear themselves of them at once but still the blood will remain dropping a great while after.[13]

Knox believed the land leech served as something akin to a mobile medic, maintaining humoral balances in indigenous people as they travelled through different climates, preventing 'the sicknesses and diseases of their neighbouring Counties' from seizing upon them.[14] On arriving home, leeches could be safely removed as the home was generally considered to pose no threat to health, making the protection of the leech unnecessary. The cultural practice of removing leeches on returning home also served as a marker, sustaining the boundary between wild and domestic, public and private, where in the one leeches were welcome and in the other not.

Lemon had long been used by those who shared their world with land leeches. It did not kill the leech nor, some believed, even deter the animal from biting. However, when applied directly it did cause leeches to stop feeding and drop off. Lemon juice was also believed to encourage healing. Odoric of Pordenone, who travelled across East Asia in the early fourteenth century, wrote of a lake full of 'horse leeches and blood suckers' where the inhabitants protect themselves with 'lemons which they peel anointing themselves with the juice'.[15] Foreign visitors to East Asia have generally been less tolerant of native leeches. Nineteenth-century colonialists, for instance, developed various forms of leech gaiter, including thigh-high boots and garments made of tightly woven cloth, leather or India rubber. Despite the fact that gaiters only encouraged leeches to climb higher up the leg and hence closer to more sensitive regions, they were preferred to the indigenous barelegged approach to daily life.[16] A century later, Gordon Stifler Seagrave, a medical missionary working in Burma, wrote of his eventual resignation to the irrepressible blood-suckers:

> Ground leeches were everywhere. I had always thought people were exaggerating when they claimed that leeches could suck a man to death, but I'm sure that many of the refugees must finally have succumbed to leeches . . . Remembering experiences in Namkhan with patients who had leeches in the nose, throat, and urethra, I finally compromised with the leeches . . . letting them get their

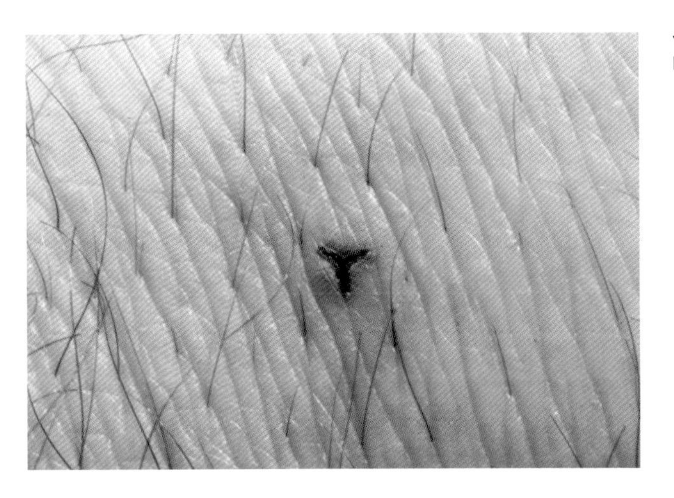

Y-shaped bite mark left by a leech.

fill so long as they kept away from my face and the fly of my trousers.[17]

TWENTIETH-CENTURY WILD LEECH

Leeches continued to mark the wild and uncivilized regions of the world throughout the twentieth century. In *Ceylon: Pearl of the East* (1963), Harry Williams expressed his exasperation with this 'obscene' creature:

> Their strength is that they are not seen or heard – they advance under cover of vegetation and their progress makes no noise at all – and having fastened themselves to their victim, man, woman, child or any kind of beast, they can slip through any protective covering. Their initial bite is not noticed, and indeed they often drink their fill of blood and roll off, distended balls, without being noticed. But if one should see them and strike them off, they leave

> their teeth behind and the result is certain to be a poisoned bite. A lighted match applied to them causes them to curl up, one hopes in agony, and depart, taking their teeth with them.[18]

Though their bite irritates and can cause infection, the leech has not generally been viewed as a global threat to health. Neither have leeches endangered food supplies in a comparable manner to creatures such as the rat. Consequently, leeches broadly escape the Western tendency to eradicate those animals it views as a threat. Whenever Western cultures have encountered wild leeches, it is generally the humans that have been 'out of place'. Land leeches are an irritant only to Westerners who travel into *their* habitats; on the whole, we have been content to let wild leeches exist over 'there' whilst we live happily over 'here'.

Twentieth-century encounters with leeches were often a consequence of military action. In the 1940s, for example, the British sought a means to repel land leeches during the Burma Campaign. No reliable repellent was developed, and the British accepted land leeches as an irritant rather than an enemy to be exterminated.[19] This attitude amazed the Americans, who admired how British and French colonialists had 'for the most part accepted leeches as part of the environment' and struggled to understand why indigenous peoples 'do not take kindly to protective or control techniques', preferring to 'accept the leech as a pest'.[20] In their fight to contain the spread of communism in East Asia, the U.S. military sponsored one the most exhaustive studies of leeches ever carried out. Though it was not known to be a communist, the U.S. declared war on the leech in the 1960s, seeking a reliable means to deter and eradicate the animal. American soldiers would not learn to 'live with' leeches, viewing them as an 'undesirable parasite'. Despite no evidence of leeches acting as disease vectors,

Werner Herzog's film *Rescue Dawn* (2007) uses wild leeches to intensify the horror of the Vietnam War.

laboratory studies were conducted to show that, in principle, leeches could carry blood-borne disease.[21] Accordingly, uniforms were strengthened and impregnated with chemicals in the effort to repel the creatures. However, soldiers found the strengthened fabric difficult to wear as it was uncomfortable in the heat and humidity of a tropical environment. Most soldiers preferred to risk leech bites rather than swelter in their secure, but restrictive, new uniforms. American expectations to be protected from leeches could not easily be balanced with the need for comfort, and breathable clothing, that was necessary to work effectively in the tropical climates inhabited by terrestrial leeches.

In Vietnam, a war that did more than most to establish the reality of post-traumatic stress disorder within American culture, the leech took its place as a cause of psychological distress:

> The attack of the leech is particularly insidious because the leech may attach itself, make its bite, and be partially engorged with blood before its presence is discovered. This may give its victim a real shock of surprise, especially if he is relatively unfamiliar with the habits of the leech. It is not surprising, therefore, that people are said to react with a

sense of panic on finding themselves the host to one or many leeches. The psychological aspect of this surprise, and perhaps panic, is usually significant.[22]

One common species, *Dinobdella ferox*, feeds only from mucous membranes, dining in the mouth, nose, throat and sometimes eyes of their host. *D. ferox* stays in residence for long periods, ranging from days to weeks. Due to their small size they are only detected when cold-like symptoms are reported, or, on occasion, when the host suffers a nosebleed or coughs up blood.[23] The 'unfavourable psychological effect' of Vietnamese leeches was widely thought to threaten troop morale, with *D. ferox* inspiring incomparable trauma in its victims.

By representing that part of nature that has not yet been tamed, leeches occupy a unique location in the American psyche. For most Americans leeches are not part of everyday life but rather something that exists in the wilds of nature, a shadow on what is left of the imagined frontier. In Stephen King's novella *The Body* (1982), for example, an encounter with leeches takes centre

North American leech (*Macrobdella decora*).

Leeches take centre stage in the iconic coming-of-age moment from *Stand by Me* (1986) based on Stephen King's novelette *The Body*.

stage, marking the crossing of temporal (youth to adulthood), existential (innocence to experience) and spatial (urban to wilderness) boundaries. Set in 1960, the story follows a transitional weekend in the lives of four thirteen-year-old boys, who embark on a journey deep into the forests of Maine to find a missing boy's corpse. Deep in the woods, seeking solace from the summer heat, the boys swim naked in a pool where they encounter ferocious 'bloodsuckers', causing them to fall into 'hysterical paroxysm' as they desperately remove the 'alien', 'dirty motherfuckin' bloodsuckers' from their bodies in 'horror'. These value-laden words speak directly to North American audiences, many of whom will have experienced similar events in their own youthful explorations of nature. One of the boys, Gordie Lachance, who narrates the story retrospectively from the present, suffers the worst: just as he thinks the episode is over he finds one final leech feeding upon his tender testicles. Lachance is forced to remove the feasting animal himself, his friends too repulsed to help, only to cause the gorged creature to burst, covering his hands and genitals in a mixture of his and the animal's blood. This episode echoes the biological transition of girl to woman, the subject of King's first published novel, *Carrie* (1974), and evokes the threat of castration, a running theme through the story. So traumatized is Lachance that decades later he cannot admit even to his wife the true origin of his crescent-shaped scrotal scar.[24] In this story the leech encounter operates as a rite of passage from youth to adulthood, innocence to worldly experience.

THREATENED WILD LEECH?

Wild leeches can also be seen as victims, serving as a means by which to 'read' the health of nature. In 1816 the physician James Rawlins Johnson lamented how the indigenous British leech, once abundant, was increasingly scarce due to

> their being more in request among medical men, and to the rapid improvements which have of late taken place in agriculture, particularly in the drainage and cultivation of wastelands.[25]

In *Into the Wild Green Yonder* (2009) in the *Futurama* series, Leela encounters an endangered Martian muck leech, which she both loves and loathes, illustrating the complex relationships between leeches and human culture.

By 1910, following a century of intensified medical consumption and agricultural 'improvements', *H. medicinalis* was declared extinct from the British Isles and thought to be endangered across Europe.[26] Rumours of the death of British leeches were, however, somewhat exaggerated. In July 1978, a German shepherd dog taken to a veterinary surgeon in Wye, Kent, was found to be carrying a leech which it had picked up whilst swimming in a flooded gravel pit at Dungeness, a site later found to house a population of *H. medicinalis*.[27] In Britain, driven by increased recognition of the significance of biodiversity and the interdependence of all forms of life, *H. medicinalis* has subsequently gained unprecedented security, protected by the Wildlife and Countryside Act of 1981 and considered by English Nature as a 'high priority' for conservation.[28] Globally, trade in *H. medicinalis* is now regulated by the Convention on International Trade in Endangered Species of Wild Fauna and Flora (CITES).[29] Furthermore, leeches now serve as 'biomarkers', or indicators of environmental quality, allowing ecologists to monitor the quality of freshwater ecosystems.[30] The presence of a healthy population of leeches, for ecologists, rather than being horrific is highly desirable.

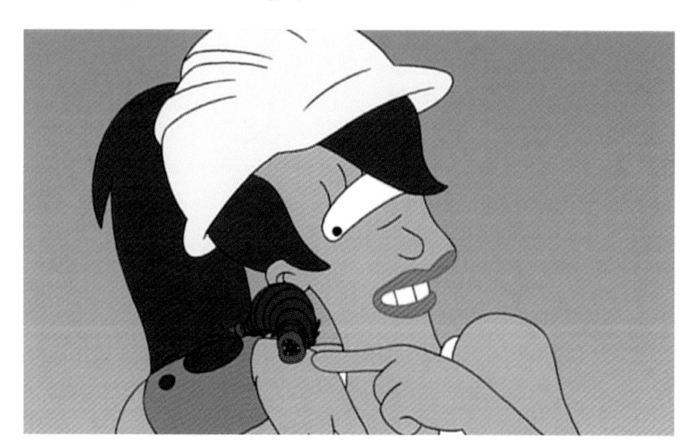

In the early environmentalist film *Frogs* (1972), leeches join nature's rebellion against human domination to devastating effect.

In 'The River' Rick Veitch introduces a mutant leech named 'Blood-sucker' whose behaviour throughout is shown to be ' not good or evil' but rather instinctual. *Eastman and Laird's Teenage Mutant Ninja Turtles*, vols 24–6 (1989).

Wild leeches possess complex identities which, far from being intrinsic to the animals themselves, emerge from the contexts in which they are encountered. The idea that leeches 'belong' in the wilderness and that 'we' belong in cities is a product of historically situated ways of seeing the world. As our planet becomes ever more densely populated, and human activity narrows the borders between wild and urban spaces, we may have to rethink our assumptions as encounters with leeches become more common.

In Japan the once rarely encountered mountain leech, *Haemadipsa zeylanica*, is now increasingly found in urban spaces. Within Japanese culture the leech enjoys a typically ambivalent status. Shinto mythology, for example, tells of the first child of Izanagi and Izanami, the central deities of creation, who was born without bones, arms or legs. Hiruko, meaning 'leech-child', was born this way because of his mother's transgression in speaking first at her marriage ceremony. Shamed, Izanagi and Izanami cast their child out to sea in a boat of reeds. Hiruko survived abandonment,

IT IS THE THING BORNE FROM THE BLOOD OF THE MUTANT TURTLE, RAPHAEL.
IT IS CALLED BLOODSUCKER...
AND IT IS HIDEOUS TO BEHOLD.
OLD MAN RIVER
STORY AND ART - RICK VEITCH LETTERS - GARY FIELDS
2

overcoming many hardships, eventually growing bones, legs and arms, and becoming the god Ebisu. Whilst Ebisu is celebrated in Shinto myth as the god of good fortune and of hardworking men, and the protector of the health of young children, leeches are less well tolerated.

The most common Japanese leech is known as *yamabiru*. These slender terrestrial leeches easily infiltrate the shoes and socks of human hosts where they can feed for up to an hour, only being noticed when they have swelled to up to ten times their original size.[31] In 2007 the Institute for Environmental Culture, a research facility located in the Chiba region east of Tokyo, reported that *yamabiru* were becoming a problem in 29 of Japan's 47 districts. But this is not simply a problem between humans and leeches. Rather, it is an effect of complex ecological interdependencies. As human rural populations migrated to the city, wild animals, including boar and deer, increased in numbers and range. Simultaneously, Japanese reforestation programmes and urban expansion narrowed the distance between wilderness and city, allowing these wild animals to enter cities, attracted by the easy sources of food. The once mountain-dwelling *yamabiru*, not wishing to be left behind, also moved into urban areas using travelling animals as something akin to 'meals on wheels'. In a globalized world, speaking in terms of *our* place and *their* place may be as spurious a way to think as that of *our* bodily fluids and *their* bodily fluids. We may do well to learn to share.

6 Horror Leech

In the twentieth-century cultural imagination, more than anything else the leech embodied horror. From the late nineteenth century, as enthusiasm for bloodletting faded, the leech began to lose its place as a trusted medical companion; at the same time, leeches took on a different form of cultural labour, becoming figurative monsters. Across popular culture, in journalism, literature, film and art, leeches came to serve as metaphors for all the worst characteristics of humanity, metaphors that mediated and perpetuated social inequalities and sustained notions of difference by encouraging fear of the other.

RACIAL LEECH

At the turn of the twentieth century long-held anti-Semitism was reignited for modernity as a result of Russian persecution, causing Eastern European Jews to migrate across Europe and to America. Within the proliferating anti-Semitic literature of the time, leeches were employed to recast Jewish identity in new and profoundly negative ways. The leech's association with blood was used to re-emphasize Jewish otherness, transferring religious difference into their very blood. In *Dracula* (1897), Bram Stoker drew on the tropes of blood and survival, foreigner and native, leech and Jew, to create the archetypal human-leech hybrid:

Count Dracula. Reinventing the vampire for an anti-Semitic age, Stoker imbued his fictional fiend with both Jewish and leech-ish stereotypes of the day. An Eastern European Count migrating to England driven by his ravenous appetite for blood becomes a foreign parasite preying on the British elite. Through the observations of his hero, Jonathan Harker, Stoker integrates the anti-Semitic imagery of the Jewish desire for wealth with the leeches' perceived insatiable appetite for blood, describing how Dracula greedily hoarded 'gold of all kinds' in Transylvania, allowing him to worm his way into the elite social circles of London.[1] In one scene, finding the Count asleep, Harker explicitly compares Dracula to a voracious leech: 'it seemed as if the whole awful creature was simply gorged with blood; he lay like a filthy leech, exhausted with his repletion.'[2] Stoker created his figure of the vampire by combining Victorian anti-Semitic paranoia (which saw Jews as parasites on nations that housed them) with the characteristics of a bloodthirsty leech. The effect was to indelibly associate leeches with human monstrosity in the modern imagination.[3]

Later, association of the Jewish people with leeches was used by Adolf Hitler in his earliest writings to construct his anti-Semitic ideology of the purity of the Aryan 'master race':

> This thinking and striving after money and power, and the feelings that go along with it, serve the purposes of the Jew who is unscrupulous in the choice of methods and pitiless in their employment. In autocratically ruled states he whines for the favour of 'His Majesty' and misuses it like a leech fastened upon the nations. In democracies he vies for the favour of the masses, cringes before the 'majesty of the people' and recognizes only the majesty of money.[4]

Leeches feature prominently in Capcom's phenomenally successful *Resident Evil* video game franchise. Professor James Marcus has created a lethal t-virus by combining a progenitor virus with leech DNA. Unexpectedly his experimental leeches learn how to work together, developing a group consciousness that allows many leeches to shape themselves into a horrific humanoid form and become a single creature, as shown here.

Dr Polidori's leech jar, from the Ken Russell film *Gothic* (1986).

Nazi propaganda mobilized leech imagery to depict Jewish people as subhuman, re-invoking ancient fears of Jewish blood drinking to ignite popular prejudice against them. The Third Reich's Ministry of Propaganda funded the film *Der Ewige Jude* (*The Eternal Jew*, 1940), promoted at the time as a documentary, claiming to show actual footage of a Jewish blood-drinking ceremony.[5] The horror of the leech and the image of the parasite enabled the Nazis to stigmatize the Jews as such, contributing to the provision of a pretext and rationale for the Holocaust.

However, the negative association of Jewishness with leeches was not limited to Nazi Germany. Rather, it was found throughout anti-Semitic thought, which for much of the twentieth century was itself prevalent across Western culture. In 1937, for instance, Winston Churchill crudely likened the stereotypical Jewish money-lender to a 'Hebrew bloodsucker', though he was, of course, later highly critical of the Nazi persecution of the Jewish people.[6] After the Holocaust, latent anti-Semitism was no longer acceptable within Western culture. Although the conflation of Jewishness with leech imagery was not forgotten, when articulated it proved highly controversial. In 2005 the American television programme *Good Morning America* aired an alleged telephone answering machine message in which the troubled King of Pop, Michael Jackson, blamed the decline in his fortunes on the Jewish people. Jackson was heard to complain of how 'they suck . . . they're like leeches . . . It's a conspiracy. The Jews do it on purpose', causing a public furore leading to demands for a formal apology.[7]

Paradoxically, despite anti-Semitic ideology having itself become monstrous within the Western imagination, the association between leeches and Jewishness continues to be invoked, even when the aim is to depict Nazi beliefs and actions as grotesque. In *Puppet Master* (1989–2004), a low-budget horror franchise, the character Andre Toulon reanimates his wife, who had been a victim of Nazi violence, in the body of a vengeful puppet known as 'Leech Woman'. To instigate her revival, Toulon inserts several leeches into the puppet's mouth. These take up residence

The Leech Woman regurgitates her leeches in *Puppet Master: The Legacy* (2003).

in her stomach, driving her thirst for revenge. The *Puppet Master* series, which to date has stretched to over ten films, constantly invokes the horrors of the Third Reich and a need for justice. Leech Woman seeks her own violent vengeance by hunting down Nazis and, when she finds them, regurgitates the leeches, which climb from her mouth to suck the blood of her victims, returning to reside within her only after the latter have been drained of life. Here the very symbols that the Nazi regime had mobilized to victimize the Jewish people have been turned back upon the oppressor. Nevertheless, despite association with the anti-hero, the leech remains ambiguously monstrous.

GENDERED LEECH

Leeches have historically been correlated with the perceived fragility of the female sex, serving as a mediator between patriarchal medical narratives that linked, for example, menstruation, blood and a propensity towards hysteria. In 1892 a hysterical New York woman attempted suicide by overdosing on leeches, attaching 50 to her body in the hope that they would drain her of blood, contributing to the association.[8] In the twentieth century these links continued to resonate as the figurative leech gave form to new ways of representing female 'weaknesses'. In 1923 *Quiver*, a popular British magazine, carried an article titled 'Human Leeches and their Devastating Ways' which extended the metaphor to describe a pathological pattern of human behaviour in highly gendered terms. The 'human leech', *Quiver* explained, mimics the natural leech, but rather than drawing on blood feeds psychically, 'extracting, if not actually living on, the vitality of others'.[9] Like their metaphorical kin, human leeches came in many forms, stealthily stalking their hosts and feeding often unnoticed. Human leeches ranged from the 'neurasthenic' type

Attack of the Giant Leeches (1959).

to the 'weak-willed' one, from the 'the flabbily sentimental' to the 'exacting and domineering'. All, noticeably, were negative caricatures of the female sex. Insecure mothers mollycoddled their children, sapping their courage and independence. Similarly, greedy and power-hungry wives dominated their husbands, shattering their confidence and ability to function in public life. Having exposed the parasitic epidemic silently infesting British households, *Quiver* asked, 'Are you a leech?' If the answer was 'yes', the article implored that the reader correct their 'human failings', which hitherto had been described only through the figure of the leech.

The idea of the psychic leech quickly embedded itself in the Western psyche, appearing, for example, in the American B-movie *Attack of the Giant Leeches* (1959). Presenting supersized leeches mutated by atomic tests, the theme of this film might first appear to be environmental pollution.[10] However, while the monstrous atomic leeches terrorize a local community near the Everglades,

CRAWLING HORROR...
RISING FROM THE DEPTHS OF HELL... TO KILL AND CONQUER!
THE GIANT LEECHES
STARRING KEN CLARK · YVETTE VICKERS · JAN SHEPARD · MICHAEL EMMET · WRITTEN BY LEO GORDON
PRODUCED BY GENE CORMAN · DIRECTED BY BERNARD L. KOWALSKI · AN AMERICAN INTERNATIONAL PICTURE

Publicity poster for Edward Dein's *The Leech Woman* (1960).

an exacting and domineering woman, Liz Walker, is shown actively depleting the strength of her husband and the life of their marriage in the manner of a psychic human leech. Parading around in black panties and bra, she refuses to allow her husband to touch her, whilst outside the marital home she becomes a notorious adulterer, reducing her husband to a local laughing-stock. This aspect of the film inspired the alternative title *She Demons of the Swamp*.

In 1960 the public appetite for depictions of the female psychic leech was fed with a new B-movie titled *The Leech Woman*, in which the parasitic relationship moved centre stage. June Talbot, the eponymous 'leech woman', embodies all the characteristics of what *Quiver* had named the 'flabbily sentimental leech'. Possessing an insatiable desire to be loved, Talbot, an ageing and neglected doctor's wife, has fallen into alcoholism and resentfulness as a result of her husband's failure to provide the affection she craves. In the course of the film June accompanies her husband

Publicity poster for *Attack of the Giant Leeches* (1959).

on a research trip to Africa, where she learns an ancient tribal ritual that allows its practitioners to restore their youth by feeding upon male pituitary fluid, physically draining the man of his vitality. Re-enacting the deep-rooted association of leeches with longevity that had previously informed the legend of the vampire, *The Leech Woman* updates this myth for a modern audience. Drawing on endocrinology, then a newly emerging medical science, the psychic leech, feeding on the mind, is reified as June Talbot is further transformed into a parasite whose life now depends on leeching the pituitary glands of men. The figure of the leech is used to convey inequalities between the sexes, as one scene of the film makes explicit: 'for a man old age has rewards' while 'the aged woman is nothing'. The leech, however, undertakes deeper metaphorical labour, literally embodying the central theme of the film. Being protandric hermaphrodites, leeches are born male, becoming female as they age; thus youthfulness in the leech world is the property of males alone, whilst female leeches are, by definition, old. In this way the leech served to focus attention upon this most difficult area of human culture, sexual equality, in a period when, on the cusp of women's liberation, relations between men and women and their attendant social roles were beginning to be fundamentally reconsidered.

SEXUAL LEECH

At its American release, *The Leech Woman* was double-billed with the British Hammer Horror Film *The Brides of Dracula*, which brought to the fore the metaphorical connections between leeches and human sexual desire. Hammer Horror films exploited the cultural connections between leeches, youth, fertility, reproduction and sex in order to recreate the vampire for an age of sexual freedom, making the creature's insatiable sexual desire equal to its voracious bloodlust. Where earlier vampires, famously Bela Lugosi, lured and paralyzed their human prey using their powers of hypnotism, Hammer transferred the action to the mouth, imbuing the vampire with a sensuous kiss that transformed the act of biting into a highly sexualized moment, with victims becoming paralysed yet apparently simultaneously aroused.[11] The mixing of bloodlust and sexual lust in the figure of the vampire was as controversial as it was lucrative. Where Hammer led, others followed.

More than most, the Canadian horror director David Cronenberg capitalized upon the linking of leeches with sexual desire. Dispensing with the anthropomorphic convention, Cronenberg brought leeches themselves to the fore. In *Shivers* (1975), leech-like parasites are created by a repressed scientist who believes humanity has lost touch with its primordial instincts. When released upon the unwitting residents of an apartment block the leeches seek out hosts, worming their way into human bodies through the mouth, anus or vagina, an experience that although portrayed as painfully bloody, for the human is also ambiguously pleasurable, reminiscent of the vampire's kiss. Once in residence, the leech

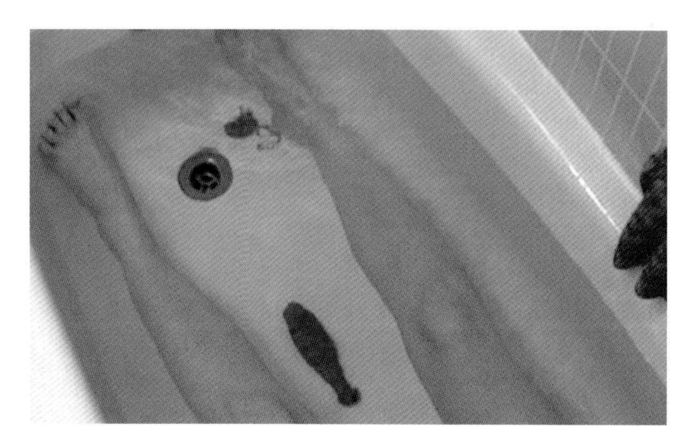

A leech-like parasite approaches a vulnerable woman taking a bath in Cronenberg's *Shivers* (1975).

works to release the host from their psychosocial repressions, freeing the resultant symbiont to realize their innermost desires. All moral restraints and taboos are forgotten and societal order collapses as individuals wildly satisfy their depraved desires, leading to violence, uninhibited sexual conduct and murder. As the alternative title, *They Came from Within*, suggests, the unrestrained debauchery that follows originates not in the state-of-the-art sterile apartment building, nor from leeches themselves, but rather from the depths of the human soul. Cronenberg himself is fascinated by leeches and kept one as a pet during the writing of the *Shivers* script, stored in a jar in his refrigerator (much to the consternation of his house-sitters, who were left precise instructions on how to feed the creature bloody pieces of raw liver, to be obtained fresh from a favoured butcher). Over 30 years later one such visitor vividly recalled how Cronenberg took great pleasure in watching his pet 'begin to strive hungrily for nourishment' when moved from the fridge into a warmer climate in anticipation of a meal.[12]

DISEASED LEECH

Blood has long been feared to contain disease. The leech, through its association with the extra-bodily circulation of blood, has similarly been framed as a vector of blood-borne disease. At the height of the nineteenth-century leech craze, many physicians were concerned that the reuse of medical leeches could spread infections such as syphilis, an ailment itself entangled with sexual promiscuity, from the morally corrupt to the innocent.[13] In the late twentieth century this history allowed the leech to be used to channel popular anxieties about blood, which, driven by the AIDS epidemic, reignited fears over the flow of blood amongst the human populace.

The film director Guillermo del Toro, for example, in his critically acclaimed *Cronos* (1993), weaved contemporary concerns over the pathological contents of blood with older themes regarding its youth-preserving properties. Central to the story is the Cronos device, created by the alchemist Fulcanelli in 1536. Reminiscent of the tempest prognosticator, the Cronos device provides a clockwork home for a bloodsucking leech, which simultaneously operates to draw blood from a human in the manner of nineteenth-century scarification. Importantly, the relationship between leech and man, mediated by machine, is not portrayed as parasitical. On the one hand, by operating the machine, the human host chooses to participate in the exchange of blood. Moreover, the relationship is symbiotic in that the leech survives on the donor's blood, whilst the donor gains in return a healthy, unlimited lifespan. The plot is driven by human competition for the Cronos device, accidently discovered by the antiques dealer Jesus Gris but sought after by Dieter de la Guardia, a wealthy man diagnosed with terminal cancer who sees the device as his only hope for a cure. Here, del Toro invokes the ancient healing power of leeches. But he does so ambiguously, for use of the Cronos device comes at a cost.

Although communing with the Cronos leech brings bodily revitalization, the relationship so produced blurs the species boundaries, as Jesus increasingly adopts leech-like behaviour, craving the blood of others. In a seminal scene set within a public toilet, Jesus, stumbling upon drops of blood accidentally left by a previous occupant who had suffered a nosebleed, cannot resist the temptation to give in to his insatiable desire for blood, bending to lick the life-giving fluid from the floor. Against the context of the moral panic about the AIDS outbreak, so recklessly exposing oneself to another's blood broke the ultimate taboo. Despite his internalization of leech-like desire, Jesus is no monster.

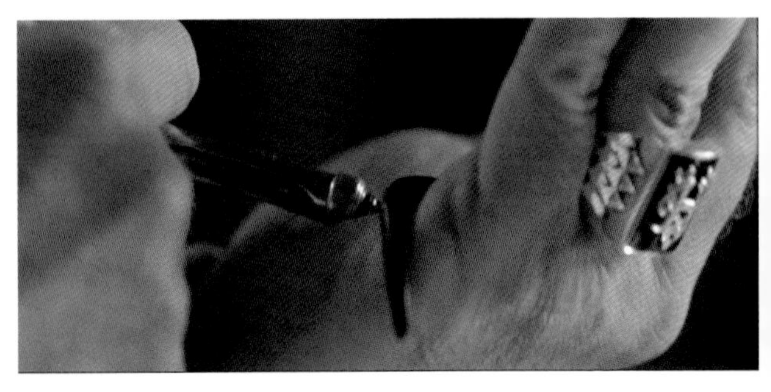

Jesus Gris licking bloody fluids off a bathroom floor in Guillermo del Toro's *Cronos* (1993).

A Reaper about to bite in Guillermo del Toro's *Blade* II (2002).

Van Helsing extracting Dracula's blood from the leech in Patrick Lussier's *Dracula 2001* (2000). The scene reflects fears about blood in the wake of AIDS.

Even after death, when transmogrified, rejuvenated and resurrected by his communion with the Cronos leech, he retains a sense of his humanity, sustained by the loving relationship of his granddaughter. In the end he chooses to abandon immortality, ending his relationship with the Cronos leech and choosing to die as he had been born, with his human kin.

In the film *Blade II* (2002) del Toro again explored the stigma and suffering caused by an apparently unstoppable blood-borne epidemic. Set in a world where human and vampire civilizations uneasily coexist, the plot revolves around a mysterious virus devastating the vampire population. Interrelating themes of evolution, degeneration and disease, the vampire-borne plague reduces the otherwise cosmopolitan bloodsucker to its savage origins. Once infected, vampires are shown to develop an uncontrollable and insatiable appetite for blood, turning on their own kind and abandoning their civilized vampire culture, fleeing from their modern urban environments to inhabit the swamp-like conditions of the sewers. As the vampires succumb to the disease their bodies increasingly exhibit leech-like characteristics: the most prominent is the development of a hideous tripartite jaw that contains not only three rows of razor-sharp teeth but also that

Hirudo medicinalis representing the shape of a leech bite.

other leech-like appendage, a feeding proboscis. Despite mobilizing the most alien physiological characteristics of leeches, the diseased vampires are portrayed not as monsters but as victims. In the final scenes of the film it is revealed that the disease came not from nature but from a biomedical experiment intended to augment vampires by allowing them to venture safely into daylight. Instead of enhancement, however, the scientific attempt to better nature had led to tragedy.

Today, though leeches have come to symbolize the horrific, they do so only as a means to reveal the dark passengers hidden deep within the human self. Distancing the leech from the human operates to separate those aspects of being human we would prefer not to remember. By drawing out and quarantining the undesirable traits of the human character, leeches, even in metaphor, perform an act of cultural cleansing, healing us of our own failings. In his films, del Toro invites us to reflect on this by raising the possibility that the gap between human and leech may not be as wide as we would like to believe. It may be that the monster, rather than being other, may lurk within.

7 Biomedical Leech

A scientific conference on the 'Biomedical Horizons of the Leech', held in Charleston, South Carolina, in 1990, opened with the following poem:

> Tonight we celebrate the leech
> A creepy-crawly sucking sneetch
> It can't stand a clotted cud
> So it anticoagulates the blood
> From Albany, Germany, Maywood and Wales
> There have come many strange, weird tales . . .

Experts from disciplines as diverse as surgery, molecular genetics, pharmacology and palaeontology had travelled across the world in the hope of 'pioneering the biomedical revival of the leech'.[1] Their aim was to bring about a leech renaissance.

It is generally thought that the use of medical leeches had all but faded by the twentieth century. However, leeches in fact retain a small but tenacious presence in medical culture even in the present day. In the early twentieth century the British appetite for leeching remained healthy, albeit much diminished since the days of leech mania. During the First World War, for example, the medical press voiced strong concern when the German invasion of France, and subsequent targeting of British shipping,

Borneo tiger leech, *Haemadipsa picta.*

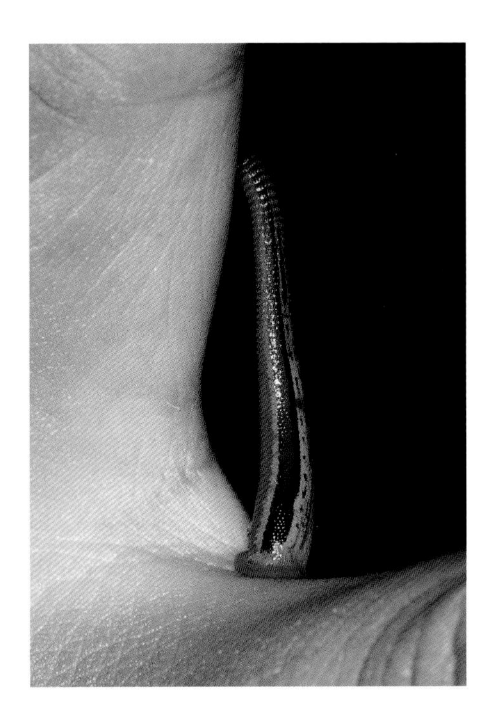

disrupted leech imports from France.[2] Leeches maintained an important role as a popular remedy for everyday problems such as gumboils, dental abscesses, headaches and black eyes into the mid-century. When imports were again disrupted during the Second World War, public alarm was aired in the popular press.[3] The use of leeches in hospitals, however, had substantially declined, driven in part by medical practitioners' desire to transform their profession. Building new alliances between medicine, science and industry led physicians to embrace symbols of progressive medical power and advancement such as new pharmaceutical drugs, in particular antibiotics. As it sought to distinguish itself from the past and establish its 'modern' credentials, scientific medicine increasingly recast the medicinal leech as an antiquated therapy.

In 1953, when news broke that the stricken Soviet premier, Joseph Stalin, was being treated with a combination of medical leeches and penicillin, the American Medical Association proclaimed that 'blood sucking worms were out of date in the United States and Europe'.[4] Against the backdrop of Cold War hostilities, the *Milwaukee Journal* asked:

> Why should Russian medicine be any different to anything else Russian? . . . Russia . . . has made tremendous industrial and technical strides – but it has retained at the same time the basic traditions and habits of uncivilized and even barbaric ages . . . Penicillin and leeches incongruous? Not to a nation which can boast at one and the same

Leeches displaying their colourful underbellies while crawling to the top of their tank, in a contemporary leech farm.

Medical leech (*Hirudo medicinalis*) feeding.

> time of some of the greatest poets and musicians, and of the world's biggest and most terrible slave labour camps, and liquidation programs.[5]

Again, leeches were mobilized to represent barbarism, inhumanity and political terror, but now for a new age. Associating leeches with the Russian Gulag made them victims of the Cold War, at least metaphorically. Appropriating leeches in this way served the political desire to claim science as the property of a superior, future-oriented and democratic West. Consequently, if only rhetorically, Western medical culture began to abandon its longest serving medical ally, paralleling, for instance, the erosion of long-held friendships in the McCarthyite anti-communist paranoia prevalent in 1950s America.

In the West, demand for *H. medicinalis* declined dramatically between the 1940s and 1970s, but did not fade entirely. During the

1950s, leeches continued to be stocked by many American pharmacists to meet public demand.[6] Similarly, across Europe leeches were still prescribed for medicinal use: at Mount Vernon Hospital in London leeching remained a common therapy to 'relieve a cardiac patient's congested liver'.[7] In Italy, medical leeching also persisted. The hotel magnate Charles Forte (1908–2007), for instance, believed he owed his life to the leeches that cured his childhood breathing difficulties, and the renowned opera singer Andrea Bocelli recalls that as a child, doctors attempted to save his eyesight by applying leeches.[8] In France, too, a small but sustained demand existed throughout the twentieth century. As a medical companion *H. medicinalis* was not, therefore, made redundant. On the contrary, leeches found new roles to play, not least in the biomedical sciences.

LEECH IN THE LABORATORY

By the turn of the twentieth century the 'laboratory revolution' had transformed medical knowledge and practice radically. Illness was now understood to be a consequence of a recently invented form of life: microbes. In the nineteenth century, pioneering scientists such as Louis Pasteur and Robert Koch famously transformed the view of disease though 'germ theory'. Some contemporaries perceived these changes as marking a transition to a new medical culture, in which 'the leech is no longer in fashion and the favour of mankind has passed to the mammalian'.[9] All manner of species, from mice, rats, guinea pigs and rabbits to cats, dogs and monkeys, now found employment in the laboratory as experimental tools for the investigation of infection and the production of vaccines and drugs.[10] New medical sciences such as bacteriology moved the site of medical innovation from the hospital to the laboratory, ushering in the

French illustration of two men sharing a joke about leeches outside a pharmacy, 20th century.

present biomedical age. Whilst leeches lost their place as the physician's medicinal companion of choice, they found new niches in the scientist's laboratory.

In 1936 the British physiologist Henry Dale and the German pharmacologist Otto Loewi were jointly awarded the Nobel Prize in Physiology or Medicine for their 'discoveries in respect of the chemical transmission of nerve impulses'. These findings, which helped to reveal how the human nervous system worked, could not have been made without laboratory leeches. At this time there was great interest in the human body's communication system, in particular as to whether it was electrical or biochemical in nature. Dale and Loewi successfully established that acetylcholine, a biochemical, served as an important neurotransmitter, crucial to the functioning of the nervous system. Because the presence of acetylcholine in the body was extremely brief, and its effects fleeting, it was very difficult to measure. Wilhelm Feldberg, a German physiologist who fled the Nazi purge of Jewish scientists to join Dale's laboratory, helped solve this problem with the help of leeches. In 1934, Dale wrote to Loewi:

> what a joy it is to have Feldberg back here, however much one deplores the conditions, which have driven him out of his own country. His importation of the leech test . . . seems likely to be as stimulating for my own work on chemical transmission, as the expulsion of the Huguenots from France was for the British textile industry.[11]

Feldberg's leech test provided a simple method to detect and measure the presence of acetylcholine. It consisted of a bath containing chemicals that would preserve acetylcholine, within which a leech muscle was suspended. Leech muscle is extremely sensitive to acetylcholine, and by measuring its contractions it could be used as way to quantify the presence of this ephemeral biochemical. Once able to detect and measure acetylcholine, Dale and Loewi could investigate how this substance operated as a neurotransmitter. This extraordinary way of working with leeches led to new treatments for human illnesses: new medicines were created to reverse the action of muscle relaxants, treat

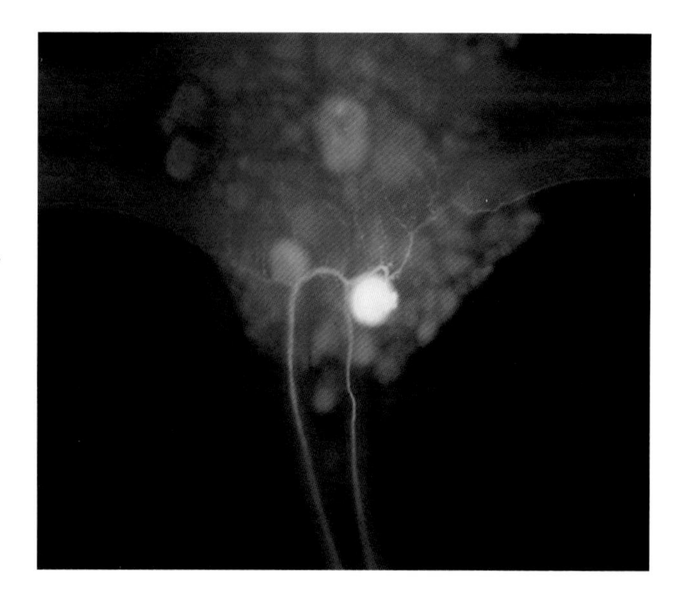

Sensory cell in a leech nerve cord ganglion. Sensory cells innervate the skin and, depending on their morphological type, respond differently to mechanical stimuli.

neuromuscular diseases (such as myasthenia gravis) and, most notably, relieve the symptoms of Alzheimer's and Parkinson's disease. All of this was possible thanks to the twitching of the leech's muscle.

Laboratory leeches have contributed more generally, too, to our understanding of the way the brain operates and communicates with the body. Humans possess highly complex nervous systems, consisting of an unfathomable number of interconnected neurons. Leeches, in contrast, possess simple nervous systems, consisting of only a relatively small number of neurons. Yet they remain capable of comparatively complex activities, making them ideal animals in which to study brain and behaviour. Leech nerve cells are located on the peripheries of their bodies, making them easily accessible, unlike human cells that are buried deep within the body. In the leech, therefore, individual cells can be identified, marked and manipulated in a way that is impossible in humans. Scientists have been attracted to the leech because they can create maps of leech neurobiology, which reveal how biology relates to overt behaviour. Ingeniously, this information can be used to investigate the biological factors that give rise to 'normal' and 'abnormal' behaviour in humans. Working with leeches, neurobiologists have also learned the incredible regenerative power of the nervous system. In one experiment, when the nervous connections that allow leeches to swim were severed, scientists were amazed when the crippled animals regained this ability within weeks. Here, the incredible healing and regenerative capacity of leeches was rearticulated using biomedicine, revealing how nervous systems could spontaneously reconnect damaged circuitry to allow animals to regain their

Muriel Robertson FRS fishing for leeches at the Lister Institute of Preventive Medicine, Elstree, *c.* 1910. Robertson used leeches to investigate the transmission of infectious diseases.

movement. By exposing the ways in which nerves develop, grow and repair themselves, leeches are teaching us not only how brains work but also how they can heal.[12] As laboratory companions, leeches give hope to those who have lost motor function following brain injury, giving promise that one day biomedicine might cure their conditions.

LEECH AS LIVING DRUG

Following a leech bite, blood flow may continue for anything up to an hour. Nineteenth-century physicians appreciated this highly unusual phenomenon, believing it made leeches superior bloodletting tools. It was only in 1884, however, that anybody thought to investigate why a leech bite differed so dramatically from other wounds. John Berry Haycraft, a British physiologist, became interested in the question when working in the famous Strasbourg laboratory of Oswald Schmiedeberg, considered to be one of the founders of modern pharmacology. Interest in coagulation was catalysed at this time because it was seen to be a vital process in stopping infection, trapping the newly discovered invading microbes whilst simultaneously preventing the wounded body from bleeding to death. Unlocking the secret of how leeches prevented the formation of blood clots would answer important questions about how the body defended itself.

Haycraft believed that leeches must introduce an active biochemical to delay the process of coagulation which, if he could find it, might explain the great mystery of why blood did not coagulate within the body, only without. To prove this biochemical existed, Haycraft dissected and processed different sections of the leech in an attempt to locate the part of the body where it was made. By placing these segments in a solution, within which the active biochemical might dissolve, Haycraft created a number of extracts that he then tested to see if they could prevent coagulation. Only one, the leech head extract, was found to delay clotting; having been concentrated by the process, it could do so for up to 24 hours. Haycraft used dogs to demonstrate that his leech extract maintained its anticoagulant power long after its removal from the leech itself. Importantly, he also showed it had no serious side effects, other than momentarily making the dogs 'a bit sad', meaning it was safe to use therapeutically.[13] Discovering a natural substance to prevent blood clotting held great promise. However, Haycraft was unable to refine his method to produce a pure extract; without this, biomedical scientists could neither study its properties nor develop it safely for clinical use.

The transformation of the leech's anticoagulant into a pure biochemical substance was achieved later by a team led by Karl Jacobj, a pupil of Schmiedeberg, at the Institute of Pharmacology in Göttingen, Germany. Between 1902 and 1905, Jacobj perfected a technique in which the heads of leeches were removed at the tenth segment, ground down and mixed with a small amount of distilled water and antiseptic before being left in cold storage for 24 hours. After this, the mixture was heated and filtered to produce a stable compound that could be used as a drug to prevent coagulation.[14] This process was patented and licensed to the pharmaceutical company Merck in 1905, which began commercially producing the new drug, naming the active ingredient hirudin.

Hirudin was immediately popular with laboratory scientists. It allowed new technologies to be developed, one innovation being the invention of the 'artificial kidney' (essentially an early dialysis machine) in 1913. At Johns Hopkins University, John Jacob Abel realized that if hirudin could be used to prevent blood coagulating, it would allow him to perfect his technique of removing blood from the body, cleansing and returning it. For the first time, a treatment for patients with failing kidneys became possible. However, the cost of hirudin was considered prohibitive at $27.50 a gram. When Abel published an article describing his artificial kidney, he advised against purchasing commercially available hirudin. Instead, he informed readers that:

> Good medicinal leeches from France can be bought in lots of one hundred or more from cupping barbers at the rate of $6 a hundred. Hynson, Westcott & Co., of Baltimore, inform us that they hope to be able to furnish the best leeches at $20 to $25 a thousand. As a single leech may yield 8 mgm. of active hirudin to extraction with water, the saving that results from making the extract is considerable.[15]

Detailed instructions on how to produce 'home-made' hirudin followed. Today, the commercial model of the pharmaceutical industry cannot be separated from the work of the biomedical sciences. That this commercial model was rejected in favour of a visit to the 'cupping barbers' is a striking example of the unpredictable ways in which disparate medical world views can be made to speak to one another. It illustrates that it is not necessary to reject old ways of doing things in order to embrace the new.

The most important interwar clinical application of hirudin was for the treatment of pulmonary embolism (where a blood clot blocks a main artery) and general thrombophlebitis (the swelling of a vein caused by a blood clot). In the interwar period, hirudin helped many physicians overcome these problems, but not without risk. Pharmaceutical companies could not mass-produce a pure drug. Consequently its use could lead to dangerous complications if the body reacted badly to contaminants in it. In 1922, a little-known French surgeon named J. Termier had a bright idea. Why, he asked, use inferior pharmaceutical products, when leeches were perfectly capable of producing and administering the pure drug themselves? Termier popularized a 'new' approach to the treatment of thrombophlebitis called the 'hirudination of the blood', which consisted of the application of leeches to the patient. In exchange for their meal, leeches gifted the patient with natural hirudin, which dissolved the problematic blood clots. Taking the drug from the leech's mouth was not only safer for the patient, who suffered no significant side effects, but the leeches were permitted to retain their heads. Termier's innovation was confirmed by numerous doctors across Europe and America.[16] The only disadvantage was said to be that:

> Leeches . . . are not pleasant to look at and create in most individuals a feeling of revulsion . . . The objection, while it may be outweighed by benefits which the method promises, is nevertheless a real one. Perhaps [one day] it will be possible to standardize a method of introducing hirudin which will replace the use of leeches.[17]

Until that day, leeches were the most reliable treatment for combating pulmonary embolisms and treating thrombophlebitis. The renaissance of hirudotherapy, however, was short-lived, because the Second World War disrupted what was left of the leech trade, cutting off supplies to European and American hospitals from southeastern Europe. By the time hostilities had ceased, a newly discovered anticoagulant called heparin had taken over the leeches' role in the West.

However, in the German Democratic Republic hirudin was not forgotten. Fritz Markwardt (1924–2011), a pharmacologist working at the University of Greifswald in East Germany,

Pharmacist at Potter & Clarks in Farringdon Street, London, holding a large bowl teeming with medical leeches, 1935.

continued to refine methods of producing the drug, consuming 15,000–30,000 animals annually. Consequently, a commercial leech industry survived in East Germany long after it had declined elsewhere. East German manufacturing techniques perfected the production of pure hirudin, though it remained time-consuming and expensive (Markwardt famously claimed that weight for weight, it was more expensive than gold[18]). Nevertheless his work sustained an interest in hirudin until advances in recombinant (synthetic) DNA technology enabled the economical production of pure hirudin, promising a completely new range of medicines. In 1997, desirudin, one of the first recombinant hirudin-based anticoagulant compounds, was released as

A drug manufactured from leeches.

a medicine, proving itself superior to heparin in the prevention of deep vein thrombosis.[19] Several other hirudin derivatives have followed, including lepirudin and bivalirudin.

Recombinant DNA techniques have relieved leeches of their role in the manufacture of hirudin-based drugs. However, hirudin is only one biochemical, produced by one species of leech. Roy T. Sawyer, an American biologist and leech expert, has described the wider leech family as a potential 'living pharmacy'.[20] Sawyer believes we should do all we can to preserve every species of leech, for we never know when we may need their help. In 1984, Sawyer established Biopharm Leeches, a modern leech farm based in Hendy, south Wales. The twenty-first-century leech farm is radically different to its predecessors. The scientific study of leeches has identified their precise needs, enabling their mass production to the highest quality. Biopharm, for instance, has developed and trademarked Hirudosalt, which when added to distilled water provides the perfect aquatic environment to keep leeches healthy. Its leeches are cultivated in vast, cold rooms, stored in specially designed tanks and maintained to the highest hygiene standards (essential given their biomedical role). Biopharm has also identified many new biochemicals produced naturally by leeches, including one that acts as a local anaesthetic (explaining why the leech bite is often painless and undetected).[21] However, the majority of Biopharm's leeches are destined to become living drugs: working to help re-establish blood flow after reconstructive surgery in hospitals across the world.

LEECH IN MODERN MEDICINE

Many of the techniques of plastic and reconstructive surgery which we might assume to be thoroughly 'modern' were in fact pioneered in the nineteenth century by the German surgeon Johann Friedrich Dieffenbach (1792–1847).[22] Some might say Dieffenbach was ahead of his time. However, he was also very much of his time, making regular use of live leeches in his surgical work. Importantly, in the nineteenth century Dieffenbach was using leeches as surgical tools, much as they are recommended today, and not for the general bloodletting more typical at the time. Twentieth-century plastic surgeons were often aware of Dieffenbach's leech use, yet were loath to imitate it. In 1929, for example, the Dutch surgeon Johannes F. S. Esser suggested that leeches could be used in plastic surgery to maintain blood flow, yet few listened.[23] In 1955 two Yugoslavian surgeons again experimented with the reintroduction of leeches, directly citing Dieffenbach's practices, but despite leeches proving to be effective, the pair would conclude: 'our aim should be to search for other methods'.[24] This general resistance to recovering leeches' use in surgical procedures persisted until the late twentieth century.

It was in France, a country that never quite lost its love of leeches, that the use of live leeches in surgery was revived. *H. medicinalis* remained a popular medicine in France, listed in the pharmacopoeia as late as 1938, whilst French pharmacies continued to stock leeches to meet public demand after this date. This consistent, albeit diminishing, demand for leeches, has allowed the original leech farm established by Bechade in 1835 to remain in business to the present day, now trading as Ricarimpex Sangsues Médicinales. In 1972 the use of leeches remained at such a level that it was necessary to have them removed from the list of therapies whose cost was covered by the French health and social security provisions. This was not a response to the leech's role in medicine per se. Rather, their removal was a consequence of new laws, reflecting wider shifts in French attitudes towards animals, which established that non-human animals should be recognized as 'sentient creatures'. Despite this, Ricarimpex continued to supply leeches to pharmacies for sale to the public until 1993! Their main customers in this period, however, were French surgeons.

By the late 1960s, advancements in microsurgery had made it possible for surgeons to reattach severed body parts, such as digits from the hands and feet. In 1972, the Bordeaux-based surgeon J. Baudet experimented with using leeches as a means to prevent blood clots and encourage post-operative blood flow. His successful techniques were widely imitated over subsequent decades.[25] In 1982, for example, a British surgeon, having observed the French practice, imported twenty leeches from Hungary and set them to work at Nottingham City Hospital. British patients, on first encountering these Eastern European medical 'assistants', were initially suspicious of this strange foreign expertise. However, once advised that leeches could make the difference between retaining and losing damaged appendages, they generally proved grateful for the animals' help. The significant problem, in the view of the chief pharmacist, was how to house leeches. Modern medical equipment was not supposed to wander about, let alone attempt to escape from the pharmacy![26] If only, one pharmacist lamented, he could locate an old leech jar.

In 1985, when Biopharm Leeches was operating from a Welsh garden shed, Sawyer received an unexpected call from Joseph Upton, a surgeon at Children's Hospital Boston in the U.S. Upton had recently reattached the ear of a five-year-old boy after it had been completely severed in a dog attack. In such cases, microsurgery usually fails because ears have such small blood vessels that it is unlikely blood flow can be re-established. This case was no different, and three days after surgery the blood clotted in the boy's ear, causing it to turn blackish-blue. Just as the ear looked lost, Upton remembered how, during his service in Vietnam, he had been forced to make use of leeches to treat wounds. Unable to locate a source of *H. medicinalis* in the U.S., Upton contacted Sawyer, who promptly dispatched a supply of leeches, reviving the Atlantic leech trade. On application of the leeches, Upton remarked that 'the ear perked up right away' and within a week normal circulation was restored. The popular media, through headlines such as 'Doctors combine modern surgery, ancient leeching to save boy's ear', marvelled at the historical peculiarities of this episode, which appeared to suggest that medicine's future lay in its own past.[27] *H. medicinalis* was propelled

back to the forefront of the public imagination. Sawyer's intuition, that medical leeches still had important roles to play in modern medicine, had been confirmed. Leeches have proved themselves superior to any other method for reviving blood flow after microsurgery: their powerful anticoagulant successfully preventing clotting, whilst their drinking of blood can be used to re-establish blood flow by drawing the blood out and into their hungry bodies.

Leeches are often the only way to re-establish blood flow after severed extremities have been surgically re-attached. Here a medical leech helps to save a patient's finger at Memorial Hospital, Rochester, New York.

In recent years postoperative hidurotharapy has saved numerous delicate body parts, bringing relief and (retrospective) gratitude to many patients. No human artifice can yet rival the leech, whose skilful suckling of fingers, thumbs, penises or breasts returns blood and life to these most culturally resonant of human body parts. Nevertheless, patients, pharmacists and nursing staff still struggle to accept the leech as a modern medical companion. It is not uncommon for leech therapy to be halted due to patient revulsion.[28] Though hirudotherapy is known to cause stress in some patients, and occasionally in onlookers such as family and nurses, these problems emerge entirely from preconceived cultural notions of what leeches are. If we could relate to leeches more as healers and less as horrors, if we recognized them as companions and not monsters, these contraindications would vanish in their entirety. After all, those who have met the medical leech have much to be grateful for. Would you not be friends for life with somebody who returned to you your functioning finger? Your healthy breast? Your working penis?

Surgeons, too, have had to reacquaint themselves with long-forgotten ways of working with *H. medicinalis*. In the 1990s, cases of 'the disappearing leech' were common as surgeons allowed leeches too much leeway in choosing where to bite. Too late, they were horrified to see the little creatures vanish deep into a patient's body.[29] These problems were discussed widely in imaginatively titled articles such as 'Taking a leech to blood: but can you make him drink?', as surgeons rediscovered the essential tools and techniques that allowed human and leech to work together.[30] In scientific medicine, however, methods have to be evidence-based. Consequently, twenty-first-century medics have found themselves creatively experimenting with leeches to discover more about their new helpers. One group of researchers, for instance, investigating ways to encourage leeches to bite, explored the animals' reactions to beer (Guinness stout and Hansa Pilsner), soured cream and garlic. Beer was found to 'disrupt the leeches' normal behaviour and made them erratic' (for which we might read 'drunk'). When the cream was smeared on glass, the leeches were found to suck enthusiastically, but when it was dabbed on to an arm, they bit no quicker with cream than without. Most curious of all, leeches seemed to be intensely attracted to garlic, but within hours of exposure to it they would die – not unlike their anthropomorphized other, the vampire.[31]

Today, only purpose-bred leeches are used medically, so as to avoid the danger of infecting a patient. In an age acutely aware that blood may act as a disease vector, leeches are never reused. Once their work is done, they are considered clinical waste. The leech cannot return to the pharmacy or be let free to enjoy retirement. Rather, they are euthanized, usually by being immersed in a solution high in alcohol, such as a mixture of ethanol and methylated spirit, before being incinerated as a biohazard. For a companion who has saved so many of the body parts we human beings prize above all others, this seems a rather sad reward.

In 2004, after an extensive evaluation of the use of leeches in medicine, alongside their production, husbandry and care, the U.S. Food and Drug Administration approved the use and marketing of *H. medicinalis* as a 'medical device'. Despite having served as a medical companion for thousands of years, today medical leeches are used as a cutting-edge biomedical technology, a living pharmacy, blurring the boundaries between the natural and the artificial. In 2004 the French Minister for Health, Philippe Douste-Blazy, was forced to justify the exclusion of *H. medicinalis* from the list of treatments supported by national health provisions. In France, he explained, leeches are recognized as 'living creatures' and so cannot be considered 'products', thus no state reimbursement can be made. This from the country

In Southeast Asia, leeches have long been associated with fertility and sexuality. In traditional Malaysian medicine, oils and fats prepared from leeches are believed to enhance male sexual prowess by strengthening the penis and preventing conditions such as erectile dysfunction and premature ejaculation.

that all but invented the leech as a tradable commodity just 200 years earlier! Revival of interest in the leech has produced much discussion about the ways in which we relate to *H. medicinalis*. One plastic surgeon, who had worked with leeches for over 30 years, emphasized how one must be 'kind to leeches' or risk having to work with 'a resentful, sullen and dispirited leech'. There is, then, no reason to discount the possibility of affective relationships between human and leech. Human relations to other animals are contingent, emerging from a complex mix of

In *X-Men: The Last Stand* (2006), as Beast extends his hand, Leech's power 'cures' the mutation.

inherited cultural, and deeply personal, contexts. Our relationships to leeches are a product of our shared histories and the stories we tell about those histories.

The recent film *X-Men: The Last Stand* (2006) illustrates many of the disparate and interwoven ways in which leeches have been related to, understood and incorporated within biomedical cultures. In the fictional universe of the X-Men, humanity has begun the next stage of evolution. Those born with the 'X-gene' possess fantastical powers and strengths, yet are viewed as abnormal by the rest of humanity. This broad theme is taken up through the character of Leech, whose own mutation allows him to completely negate the effects of the X-gene in other 'mutants'. In human hands, Leech becomes the source of 'the cure', a biochemical derived from his DNA, isolated by biomedical science and turned into a product by a pharmaceutical company. 'The cure' is shown to permanently 'correct' the X-gene, turning 'mutant' into 'human'. Leech is subsequently imprisoned, being the only source of the cure, which the pharmaceutical company proceeds to offer on a voluntary basis to those mutants who wish to take it. At the same time, the drug is weaponized on behalf of the state in preparation for an expected war between humans and mutants. Some mutants, such as Rogue, relate to the drug as a cure, choosing to take it and become 'human'. Others see the drug, and Leech, as a threat. The mutant Magneto, a Holocaust survivor who cannot escape his past, believing that humans will inevitably wish to eradicate mutants due to their inherent fear of difference, leads a band of mutants to kill Leech before the drug 'kills' them by erasing their mutant identity. A further group of mutants, the heroic X-Men, seek to protect Leech. Despite the danger he poses, they recognize his nature is no more his fault than their own is theirs. For the X-Men, the only way forward is for all involved to find a way to live together.

Through the character of Leech, a horror to some, a healer to others, *X-Men: The Last Stand* illustrates the way that relationships shape how we view the essence of another. This story deliberately evokes memories of eugenics and state sterilization programmes to reveal how fears of difference can lead to tragedy and genocide. For some, Leech is an abomination, tolerated only as long as his body is needed for the eradication of yet worse horrors. For others, Leech is the cure, bringing relief from suffering and the abnormal. For yet others, Leech is a boy in need of protection. But Leech is just an innocent, caught in a story to which he consciously contributes nothing at all.

Concluding Leech

> You have made your way from worm to human, and much in you is still worm.
>
> Parasite: that is a worm, a crawling withering worm that wants to glut itself on your infested nicks and niches.
>
> Friedrich Wilhelm Nietzsche, *Thus Spoke Zarathustra* (1883–5)

Sung in the style of a popular schoolyard rhyme, Emilie Autumn's 'Miss Lucy Had Some Leeches' provides a commentary on modern medicine.[1] Autumn, a pioneer of the 'Victoriandustrial' musical genre, draws directly from nineteenth-century culture to share her experience as a woman with bipolar disorder navigating early twenty-first-century society. Leeches feature prominently in her accounts of psychiatric illness. By drawing vivid contrasts between, for example, present-day practices of self-harming to relieve internal tensions, and past beliefs in the leech's capacity to relieve hysteria through bloodletting, Autumn mobilizes the leech to destabilize our assumptions about 'normal' and 'abnormal' mental states. Why should cutting be considered a pathological form of self-harm at one place in one time, yet leech therapy be considered curative at another? Is there an answer to be found less in the character of the participants themselves, than in the power relations between them and the medical profession that labels their 'condition'? In Autumn's hands leeches work to remind us of the historical contingency of medical beliefs. In her music she uses leeches to symbolize an overly patriarchal medicine, which continues to define and determine women's bodies, often without their full consent. Her work is all the more compelling for its ambivalent representation of the animal: within Autumn's oeuvre, leeches capture the darker side of medicine

Emilie Autumn uses leeches to embody and humanize mental illness in her semi-autobiographical *The Asylum for Wayward Victorian Girls* (2010).

MANIC
DEPRESSION
SCHIZ o PHRENIA

whilst simultaneously being anthropomorphically portrayed, appearing with large human-like eyes as a figure to befriend, not fear. By drawing on the leech's ambiguity, Autumn, amongst other twenty-first-century artists, has embraced the leech as a literary and figurative companion, transposed from the medical past to help understand our biomedical present.

Kira O'Reilly is a performance artist whose body is her medium. Her work probes the ways in which biomedicine has become the predominant means by which we understand and represent our own bodies in the contemporary world. In *Bad Humours / Affected* (first presented in 1998), O'Reilly performs with two leeches, replaying the nineteenth-century bloodletting encounter as a means to explore the power of modern biomedicine. The piece begins with O'Reilly kneeling on the floor, facing down, her upper body naked and folded in upon itself so as to shield her breasts. Her lower body is clothed in a long, flowing white skirt. An assistant places two leeches upon her pale, bare back, allowing them to choose their place to bite, rest and feed. Over the following hour the animals gradually expand, extending their bodies as they take sustenance from O'Reilly. Once satiated they choose to detach themselves, rolling down into the folds of the skirt, to be rescued by the assistant and placed in a glass leech jar in view of the audience, still very much part of the performance. O'Reilly remains at rest, continuing the intimate exchange of bodily fluids by permitting the gift of hirudin to circulate through her now open and bleeding body. Her deep red blood flows first down her pale back, then into and over her white skirt, creating a vivid contrast.[2]

Bad Humours / Affected takes a means of encountering leeches from the past and deploys it as a commentary on the present. The material intimacy of the encounter, human skin upon leech skin, together with the exchange of bodily fluids, reinforces the sense in which this is a collective work, a shared experience and a collaboration across species boundaries. Whilst O'Reilly provides her body to shape the performance environment, the leeches use their own determination to choose where to bite, how to move whilst feeding and when to complete their contribution. In this work the leech becomes a performance artist, and the penetration of the body is rescued from the realm of the taboo and re-valued. In communion with her leeches, O'Reilly presents a critical engagement with contemporary understandings of the body and its relations to other, not necessarily human, bodies. The work aims to explore and rupture biomedical power, yet it cannot avoid being woven into the very fabric of that which it seeks to destabilize. For instance, prior to one performance, O'Reilly was asked to supply an assessment of her psychiatric state, previous psychiatric and psychological history, and information regarding her Hepatitis B, Hepatitis C and HIV status.[3] No such enquiry was made of the leeches. Through her acute historical awareness, O'Reilly knowingly positions herself within, yet without, biomedical worlds, being part of, yet critically resisting, the ways in which biomedical power fashions and subjugates twenty-first-century bodies. In collaboration with leeches, O'Reilly creates a performance in which the leeches' piercing of her skin, and opening up of her body, channels history to similarly open up biomedicine, inviting her audience to question the extent to which we should allow biomedicine to determine our worlds.

The Welsh artist Kathryn Ashill invites her audience to think more positively about the relationships we might share with leeches. In her performance piece *Love Bite* (2008), Ashill brings to the fore themes of social interaction, material interventions, contact and touch. Ashill spent two weeks at Biopharm Leeches learning to breed, care for and dispatch *H. medicinalis* to customers all over the world. *Love Bite* plays with the ideal of motherhood

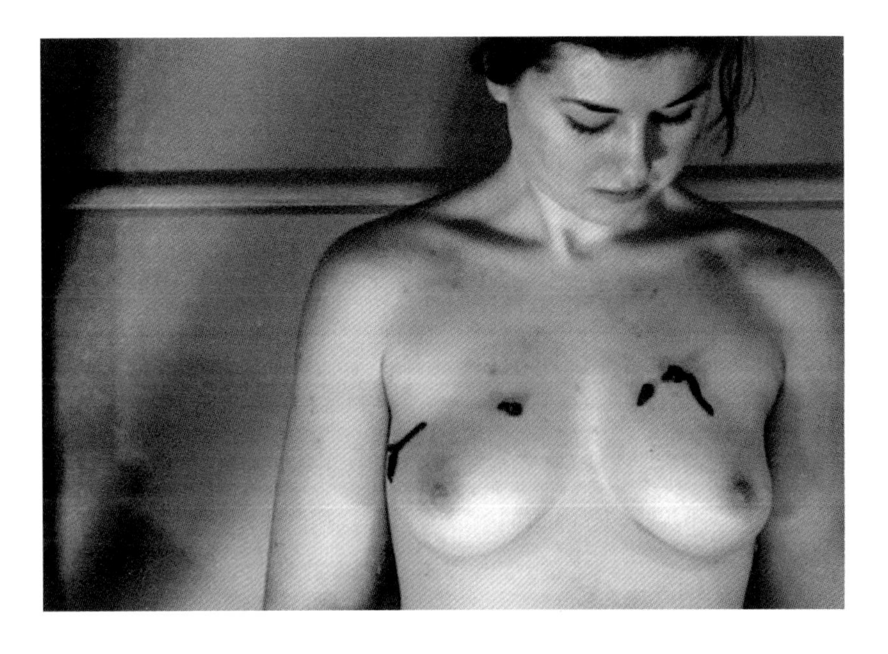

and maternal feeling, depicting leeches, which Ashill has cared for since they hatched from cocoons, feeding from her breasts. *Love Bite* emphasizes the positive role leeches play in modern medicine, recalling for instance how leeches show great care for the human breast (amongst other body parts) by reviving blood flow after reconstructive surgery. In the performance, Ashill invites leeches to be included within the human family. Insisting that the leech's oral intercourse is a 'loving bite that reinvigorates' and 'brings back to life', her communion with leeches attempts to relate apparently incommensurable worlds to one another. In so doing, Ashill and her leeches seek to open up possibilities for new ways of living and being together.

Kathryn Ashill, *Love Bite* (2008).

Throughout this book we have deliberately resisted classifying leeches as parasites. Naming another *ascribes* character as

Kira O'Reilly performing *Bad Humours / Affected* (1998).

much as it *describes* character. It should not be undertaken lightly. The name 'parasite' is already laden with values, and values can do dangerous work. Labelling leeches as parasites legitimates human disgust for these creatures. Having othered the leech, it is then a simple step to other groups of humans, who are made to take on this name. As Theodor Adorno wrote, 'what was not seen as human and yet is human, is made a thing'. And things can be treated inhumanely.[4] In the darkest moments of our shared history, millions have died because some have successfully labelled others as parasites. Why would we wish to use this term? To name another as parasite is not to name their character. It is to ascribe a negative value to the relationship one has with that other. It is to judge. It is, moreover, to interrupt a relationship, and say it should not be. The difference between a parasitic and symbiotic relationship is easily understood with reference to value, but more complex to define in practice. A parasite takes but does not give. Symbionts live together, depending on each other for their existence. To call leeches parasites is unwarranted, unjustifiable and, perhaps worst of all, dangerously arrogant.

The French philosopher Michel Serres uses the notion of the parasite to question human exceptionalism, revealing how human identity is always already entangled with that of non-humans. As humans, he writes, 'we parasite each other and live amidst parasites . . . they constitute our environment'.[5] It is *we* that are parasites. Those of us who aspire to comment upon the world – scientists, philosophers, historians, cultural critics, academics and writers of all species – take what we critique and consume it: 'the parasite one parasites the parasite'.[6] All life is interconnected, all life relates. No life can escape the relations of power it is born into. But, power relations can be reshaped. This is the possibility Nietzsche explores when Zarathustra asks the 'conscientious one' whether he knows the leech 'to its ultimate basis'. Nietzsche asks whether parasitism is in fact the human condition itself, shaping human and non-human relations alike. Is not all human knowledge built upon our own parasitism? Has not the person who studies leeches used them without giving back to them? Nietzsche throws humanity in with the worms, all sharing that lowest form of relationship: parasitism. On first reading this is a chilling denunciation of what it is to be human. Yet, by erasing difference between human and non-human, it also opens up the possibility for positive change. We should refuse, however, Nietzsche's evolutionarily conceived path to be *better than* human. We should resist, even, to try to become *better than* worms. Instead, our path should be to seek to live better *with* worms. By releasing the leech from its role as parasite we begin the work of releasing ourselves from the same. Leeches need only be parasites as long as we ask them to contain and embody our darkest nature. This strategy to make humanity virtuous succeeds only in making us parasites. Further, it establishes a cultural parasitism that forces leeches to become the symbol of their own oppression. The film *Inception* (2010) asks: 'what is the most resilient parasite? Bacteria? A virus? A . . . worm?', and answers: 'a single idea from the human mind'. Such an idea 'can transform the world and rewrite all the rules'. If there is a moral behind our leech stories, it is that by neglecting our shared history we lose our ability to understand ourselves and our relationships to others. We forget that we have *made* these relationships and therefore can *change* them. Instead of writing the rules, we let ourselves be written and determined by them. Let us not be imprisoned by this fate. Let us not be parasites. Let us refuse to see others as parasites. Let us work to find new ways to live together. Let us learn to love the leech!

Timeline of the Leech

544 MYA

(Cambrian period) First annelids emerge

1.5 MYA

(Jurassic period) First leeches are believed to have evolved alongside the earliest mammals

***c*. 1900–1500 BC**

(Old Babylonia) Earliest recorded use of leeches for medicinal bloodletting

***c*. 1567–1308 BC**

(Eighteenth dynasty of ancient Egypt) Sepulchre wall paintings of pharaohs show leeches being used for medicinal purposes

***c*. 980–1027 AD**

Avicenna includes leeches within a systematic account of humoral medicine, providing detailed instructions on their selection, care and use

1817

The French zoologist Georges Cuvier defines leeches as worms possessing 'red blood'

***c*. 1820s**

At the height of the 'leech craze' millions of leeches were transported around the world to satisfy demand from the medical profession

1827

Darwin presents his first scientific work in Edinburgh, demonstrating that miniscule black peppercorn-like creatures found in oyster shells are skate leeches (*Pontobdella muricata*)

1835

Colossal fortunes begin to be made from breeding leeches, for example M. Béchade in France

1910

After a century of over-harvesting and agricultural change, *Hirudo medicinalis* is (prematurely) declared extinct in the British Isles

1930

Max Ernst's *A Little Girl Dreams of Taking the Veil* includes leeches in his Surrealist exploration of religious ecstasy and erotic desire

1937

The International Leech Centre is established in Udelnaya, outside Moscow. Today 3 million leeches are produced annually

1758

In the tenth edition of his magnum opus *Systema Naturae*, Linnaeus classifies leeches with the *vermes* (worms)

1799

President George Washington is treated on his deathbed with leeches, possibly draining the life out of him

1809

Jean-Baptiste Lamarck, the French naturalist and early evolutionist, separates the annelids or 'ringed worms', including leeches, from the wider family group of worms

1816

French revolutionary physician François-Joseph-Victor Broussais publishes *Examen de la doctrine médicale généralement adoptée*, making blood-letting and the leech the central treatment for all illness

1851

George Merryweather draws huge crowds at the Great Exhibition in London to witness how his leeches accurately predict the weather

1884

British physiologist John Berry Haycraft discovers hirudin, a powerful biochemical produced by leeches that prevents the coagulation of blood

1897

Bram Stoker hybridizes human and leech to create the iconic monstrous bloodsucker, Count Dracula

1959

The B-movie *Attack of the Giant Leeches* provides an outlet for American enthusiasm for monstrous leeches

1987

International trade in *H. medicinalis* is restricted under the Convention on International Trade in Endangered Species (CITES)

2001

Scientists at the University of Wisconsin invent an artificial leech for the 21st century. At the time of writing, it is no more successful than previous attempts to replace living leeches in medicine

2004

H. medicinalis is approved as a safe 'medical device' by the Food and Drug Administration of the U.S. government

References

1 NATURAL LEECH

1 Charles Darwin, *The Formation of Vegetable Mould Through the Action of Worms* (London, 1882), p. 316.
2 Jean-Baptiste Lamarck, *Histoire naturelle des animaux sans vertèbres* (Paris, 1818), vol. v, pp. 290–92.
3 Robert Hooper and John Quincy, *A New Medical Dictionary* (Philadelphia, PA, 1817), p. 433.
4 Edward E. Ruppert, Richard S. Fox and Robert D. Barnes, *Invertebrate Zoology: A Functional Evolutionary Approach* (London, 2004), p. 421.
5 Georges Cuvier, *Le Règne animal distribué d' après son organisation* . . . (Paris, 1817), vol. II, p. 531.
6 Mark E. Siddall, Elizabeth Borda and Gregory W. Rouse, 'Toward a Tree of Life for Annelida', in *Assembling the Tree of Life*, ed. Joel Cracraft and Michael J. Donoghue (Oxford, 2004), pp. 237–51.
7 James Rawlins Johnson, *A Treatise on the Medicinal Leech: Including its Medical and Natural History, with a Description of its Anatomical Structure: also, Remarks Upon the Diseases, Preservation and Management of Leeches* (London, 1816), pp. 95–7.
8 E. P. Evans, *The Criminal Prosecution and Capital Punishment of Animals* (London, 1987), p. 27.
9 R. T. Sawyer and K. Dierst-Davies, 'Observations on the Physiology and Phylogeny of Colour Change in Marine and Freshwater Leeches (Annelida: Hirudinea)', *Hydrobiologia*, 44 (1972), pp. 215–36.

10 Mark E. Siddall, 'Leech Evolution', in *McGraw-Hill Yearbook of Science and Technology 2003* (New York, 2003), pp. 218–21.
11 Mark E. Siddall, 'The Beauty of Leeches', *Center for Biodiversity and Conservation Newsletter* (Fall 2000), p. 1.
12 Barnaby Conrad, *Famous Last Words* (London, 1962), p. 78.

2 MEDICAL LEECH

1 Nathan Wasserman, 'On Leeches, Dogs and Gods in Old Babylonian Medical Incantations', *Revue d'assyriologie et d'archéologie orientale*, CII (2008), pp. 71–88.
2 Stanley Finger, *Minds Behind the Brain: A History of the Pioneers and their Discoveries* (Oxford, 2005), p. 15.
3 Alfred Forke, *Lun-Heng: Philososphical Essays of Wang Chung* (London, 1907), pp. 156–8.
4 Dr Ryan, 'Lectures', *London Medical and Surgical Journal*, VII (1836), P. 200.
5 Henry E. Sigerist, *A History of Medicine, Volume II: Early Greek, Hindu and Persian Medicine* (Oxford, 1961), p. 30.
6 James Rawlins Johnson, *A Treatise on the Medicinal Leech . . .* (London, 1816), p. 3.
7 Pliny the Elder, *Natural History*, ed. John Healey (Harmondsworth, 1991), p. 113.
8 O. Cameron Gruner, ed., *The Canon of Medicine of Avicenna* (London, 1930), pp. 512–14.
9 Barbara S. Bowers, *The Medieval Hospital and Medical Practice* (London, 2007), esp. pp. 176–183.
10 Jon Cannon, *Cathedral: The Great English Cathedrals and the World that Made Them* (London, 2007), p. 194.
11 F.-J.-V. Broussais, *Leçons du docteur Broussais, sur les phlegmasies gastriques, dites fièvres continues essentielles des auteurs, et sur les phlegmasies cutanées aiguës* (Paris, 1819).
12 Jean-François Braunstein, *Broussais et le matérialisme: médecine et philosophie au XIXe siècle* (Paris, 1986).
13 J. D. Rolleston, 'F.J.V. Broussais, 1772–1838: His Life and

Doctrines', *Proceedings of the Royal Society of Medicine*, XXXII (1939), p. 408.

14 Anne Mortimer Young, 'Bleeding Antiques Part 3: Leeching', *Medical Collectors Association Newsletter*, 11 (1987), p. 1.

15 Roy T. Sawyer, 'Why We Need to Save the Medicinal Leech', *Oryx*, XVI (1981), pp. 165–8.

16 Erwin H. Ackerknecht, *Medicine at the Paris Hospital, 1794–1848* (Baltimore, MD, 1967).

17 Frederick Alexander Simon, *Der Vampirismus im neunzehnten Jahrhundert* (Hamburg, 1830).

18 B. W. Payton, 'Medicinal Leeches: The Golden Age', *Bulletin of the Canadian Historic Association*, I (1984), pp. 79–90.

19 'On Fashion in Physic', *London Magazine*, 13 (1825), pp. 177–91, pp. 189–90.

20 'Hirudiculture', *Journal of Agriculture*, VIII (1859), pp. 641–8, p. 642.

21 Johnson, *Treatise on the Medicinal Leech*, pp. 51–63.

22 Rees Price, *A Treatise on the Utility of Sangui-Suction, or Leech Bleeding: In the Treatment of a Great Variety of Disease* (London, 1822), p. 128.

23 'Editor's note', *Lancet*, VIII (1827–8), p. 4.

24 Price, *Treatise on the Utility of Sangui-Suction*, p. 94.

25 A. O., 'Leeches Drunk Will Bite Until Sober', *The Chemist*, 1 (1849–50), pp. 231–2.

26 Jonathon Osborne, 'Observations on Local Bloodletting and Methods of Practicing it', *Dublin Journal of Medical and Chemical Science*, III (1833), pp. 334–42.

27 Edward John Tilt, *A Handbook of Uterine Therapeutics and of Diseases of Women* (London, 1881), p. 136.

28 *The Household Encyclopaedia, or Family Dictionary, Volume II* (London, 1859), pp. 133–5.

29 George Horn, *An Entire New Treatise on Leeches* (London, 1798), p. 29.

30 John Lord Campbell, *The Lives of the Lord Chancellors* (London, 1847), pp. 617–18.

1 George Horn, *An Entire New Treatise on Leeches* (London, 1798), pp. 7–8.
2 John Blake, *A Family-Text Book for the Country; or, The Farmer at Home* (New York, 1857), p. 244.
3 'The Propagation of the Leech in France', *Lancet*, XXX/778 (28 July 1838), p. 621.
4 Rees Price, *A Treatise on the Utility of Sangui-Suction, or Leech Bleeding: In the Treatment of a Great Variety of Disease* (London, 1822), p. 130.
5 'On the Traffic with Leeches', *Boston Medical Intelligencer*, 4 (1826–7), pp. 234–5.
6 'Notices to Correspondents', *Australian Medical Journal*, XV (1870), p. 228.
7 'The Boundary Question', *The Standard* (10 August 1840), p. 2.
8 'The Leech: An Interview with a Phlebotomist', *St Louis Globe-Democrat* (16 January 1876), p. 12.
9 'The Leech Establishment at Smyrna', *Ipswich Journal* (24 January 1852), p. 4.
10 Thomas Brightwell, 'On the Medicinal Leech', *Provincial Medical and Surgical Journal*, X/36 (1846), pp. 428–30.
11 Richard Rathbun, 'The Leech Industry', in *The Fisheries and Fishery Industries of the United States* (Washington, 1887), vol. II, pt 22, pp. 811–15.
12 'The Leech: An Interview with a Phlebotomist', p. 12.
13 Kathleen Stokker, *Remedies and Rituals: Folk Medicine in Norway and the New Land* (Minnesota, MN, 2007) p. 215.
14 'The Two Ends of the Horse Leech', *Penny Satirist* (19 March 1842), p. 2.
15 'The Cry of the Horse Leech', *Reynold's Newspaper* (28 June 1857), p. 8.
16 'The Leech Tribes', *Bristol Mercury* (30 August 1862), p. 3.
17 J. M. Rymer, *Varney, the Vampire; or, The Feast of Blood: A Romance* (New York, 1973), vol. I, p. 507.
18 Karl Marx, *Capital: Critique of Political Economy, Volume I*

(New York, 2007), p. 282.

19 Queen, 'Death on Two Legs (dedicated to . . .)', *A Night at the Opera* (EMI, 1979).

20 Ioannes Somp, *Sanguisugae: Dissertatio Inauguralis Physiographico-Medica* (Budapest, 1843).

21 Roy T. Sawyer, 'The Trade in Medicinal Leeches in the Southern Indian Ocean in the Nineteenth Century', *Medical History*, XLIII (1999), pp. 241–5.

22 Margareta Modig, 'The Strange Lore of Leeches', *Pharmacy in History*, XXVIII (1986), pp. 99–102.

23 Roy T. Sawyer, 'Why We Need to Save the Medicinal Leech', *Oryx*, XVI (1981), pp. 165–8.

24 'The Tax Upon Imported Leeches', *Lancet* (1840), p. 319.

25 Modig, 'The Strange Lore of Leeches', p. 100.

26 M. Guibourt, 'On the Means of Increasing the Number of Leeches', *British and Foreign Medical Review*, 1 (1835), pp. 281–2.

27 Jonathan Pereira, *The Elements of Materia Medica and Therapeutics, Volume II* (London, 1857), pp. 733–4.

28 'Hirudoculture', *Journal de Chimie Médicale*, 4 (1856), pp. 174–85.

29 Robert P. Negus, *Essay on Leeches: A Practical Hand Book* (Melbourne, 1868).

30 'France', *New York Times* (1 April 1854), p. 2.

31 'Scientific and Useful', *The Queenslander* (18 April 1868), p. 3.

32 'The Leech Trade of Australia is Becoming an Important Business', *Milwaukee Daily Sentinel* (7 March 1867), p. 1.

33 'Leeches for London', *British Medical Journal* (1867), p. 274.

34 'The Week', *British Medical Journal* (1863), p. 645.

35 'The Horrors of Leech Breeding', *Rocky Mountain News* (2 February 1896), p. 18.

36 'Driving Horses Out of a Leech Swamp', *Frank Leslie's Illustrated Newspaper* (16 June 1866), p. 204–5.

37 'The Leech: An Interview with a Phlebotomist', p. 12.

1 John Baron, *The Life of Edward Jenner* (London, 1838), vol. I, p. 23.
2 Edward Jenner, 'Observations of the Natural History of the Cuckoo', *Philosophical Transactions of the Royal* Society, LXXVIII (1788), pp. 219–37.
3 Vladimir Janković, *Reading the Skies: A Cultural History of English Weather, 1650–1820* (Manchester, 2000).
4 'On the Prognostics of Leeches', *Monthly Magazine and British Register*, XXI (1806), p. 219.
5 Robert Southey, ed., *The Life and Works of William Cowper* (London, 1836), vol. VI, p. 82.
6 George Merryweather, *An Essay Explanatory of the Tempest Prognosticator in the Building of the Great Exhibition for the Works of Industry of All Nations* (London, 1851), p. 60.
7 Ibid., p. 47.
8 Ibid., p. 59.
9 Ibid., p. 44.
10 Records retained by the Whitby Literary and Philosophical Society, Whitby Museum, North Yorkshire.
11 Katherine Anderson, *Predicting the Weather: Victorians and the Science of Meteorology* (Chicago, IL, 2005).
12 'The Surprising Tempest Prognosticator', *Observer* (24 March 1851), p. 6.
13 Isobel Dixon, *The Tempest Prognosticator* (London, 2011).
14 'Preservation of Leeches', *The Medico-Chirurgical Review*, 21 (1834), p. 230.
15 Andrew H. Smith, 'An Artificial Leech', *Medical Record*, 4 (1869–70), pp. 406–7.
16 Teunis Willem Van Heiningen, 'Jean-Baptiste Sarlandière's Mechanical Leeches (1817–1825): An Early Response in the Netherlands to a Shortage of Leeches', *Medical History*, LIII (2009), pp. 250–70.
17 L. Wecker, 'De la sangsue artificielle (modéle du baron Heurteloup), et de son emploi dans le traitment des maladies des

yeux', *Bulletin Général de Thérapeutique Médicale et Chirurgicale*, LXII (1862), pp. 107–16.

18 Smith, 'An Artificial Leech', pp. 406–7.

19 *Official Descriptive and Illustrated Catalogue of the Great Exhibition of the Works of Industry of All Nations* (London, 1851), vol. I, p. 465.

20 'The Leech: An Interview with a Phlebotomist', *St Louis Globe-Democrat* (16 January 1876), p. 12.

5 WILD LEECH

1 William Dalton, *Lost in Ceylon: The Story of a Boy and Girl's Adventures* (London, 1861), p. 89.

2 J. D. Hooker, *Himalayan Journals* (London, 1854), vol. II, p. 17.

3 Ibid., p. 167.

4 Ernst Haeckel, *A Visit to Ceylon* (London, 1883), p. 138.

5 Tim Mackintosh Smith, ed., *The Travels of Ibn Battutah* (Basingstoke, 2003), p. 247.

6 Henry Marshall, *Ceylon* (London, 1846), p. 15.

7 James Emerson Tennent, *Ceylon: An Account of the Island, Physical, Historical and Topographical* (London, 1860), pp. 221, 301–2.

8 Ibid., pp. 302–7.

9 C. F. Gordon Cumming, *Two Happy Years in Ceylon* (Edinburgh, 1892), vol. I, p. 116.

10 Haeckel, *A Visit to Ceylon*, p. 139.

11 Edgar Leopold Layard, 'Rambles in Ceylon', *Annals and Magazine of Natural History*, 11 (1853), pp. 224–36.

12 Emerson Tennent, *Ceylon*, p. 307.

13 Robert Knox, *An Historical Relation of the Island Ceylon, in the East Indies* (London, 1681), p. 25.

14 J.H.O. Paulusz, ed., *An Historical Relation of the Island Ceylon by Robert Knox* (Dehiwala, 1989), vol. I, p. 7.

15 'The Journal of Friar Odoric', in John Mandeville, *The Travels of Sir John Mandeville* (London, 1915), p. 339.

16 Henry Charles Sirr, *Ceylon and the Cingalese* (London, 1850), vol. I, p. 208.

17 Gordon Stifler Seagrave, *Burma Surgeon Returns* (New York, 1946), pp. 25–6, 61.
18 Harry Williams, *Ceylon: Pearl of the East* (London, 1963), p. 238.
19 F.M.G. Stammers, 'Observations on the behaviour of land leeches (Experiments carried out on *H. zeylanica* at St. Bartholomew's Hospital in London in 1944 at the request of the Army)', *Parasitology*, XL (1950), pp. 237–46.
20 Walter Veazie and Reynolds Overbeck, *Research Report: State of the Art Study on Leech Repellents* (Ohio, 1963).
21 R. E. Shope, 'The Leech as a Potential Virus Reservoir', *Journal of Experimental Medicine*, CV (1957), pp. 373–82.
22 Veazie and Overbeck, *Research Report*, p. 2
23 H. L. Keegan, M. G. Radke and D. A. Murphy, 'Nasal Leech Infection in Man', *American Journal of Tropical Medical Hygiene*, XIX (1970), pp. 1,029–30.
24 Ibid., p. 526.
25 James Rawlins Johnson, *A Treatise on the Medicinal Leech* (London, 1816), p. 41.
26 W. A. Harding, 'A Revision of the British Leeches', *Parasitology*, III (1910), pp. 130–201.
27 Philip Wilkin, *A Study of the Medicinal Leech, 'Hirudo medicinalis', with a Strategy for its Conservation*, (London, 1987) p. 18.
28 Roy T. Sawyer, 'Why We Need to Save the Medicinal Leech', *Oryx*, XVI (1981), pp. 165–8; J. A. Bass, 'Species Action Plan: Medicinal Leech *Hirudo medicinalis*', Natural Environmental Research Council, 1996.
29 See www.cites.org.
30 J. L. Metcalfe, M. E. Fox and J. H. Carey, 'Freshwater Leeches (Hirudinea) as a Screening Tool for Detecting Organic Contaminants in the Environment', *Environmental Monitoring and Assessment*, 11 (1988), pp. 147–9.
31 Reuters Tokyo, 'Leech Invasion Makes Residents See Red' (6 September 2007) www.reuters.com.

6 HORROR LEECH

1 Bram Stoker, *Dracula* [1897] (Harmondsworth, 2003), p. 55.
2 Ibid., p. 60.
3 Nina Auerback, *Our Vampires, Ourselves* (Chicago, IL, 1995).
4 Eberhard Jäckel, ed., *Hitler: Sämtliche Aufzeichnungen, 1905–1924* (Stuttgart, 1980), pp. 88–90.
5 Eric Rentschler, *The Ministry of Illusion: Nazi Cinema and its Afterlife* (Boston, MA, 1996), pp. 160–65.
6 'Uncovered, Churchill's Warnings about the "Hebrew Bloodsuckers"', *Independent on Sunday* (11 March 2007), p. 8.
7 'Jackson "Has Anti-Semitic Streak"', http://news.bbc.co.uk (24 November 2005).
8 'Note', *New York Medical Journal*, LVI (1892), p. 103.
9 Mona Maxwell, 'Human Leeches – and their Devastating Ways, Real-Life Vampires', *Quiver* (February 1923), pp. 367–70.
10 Darryl Jones, *Horror: A Thematic History in Fiction and Film* (New York, 2002).
11 Peter Hutchings, *Hammer and Beyond: The British Horror Film*, (Manchester, 1993).
12 Bill Gladstone, 'An Encounter with David Cronenberg' (1 December 2011), available at www.billgladstone.ca.
13 'Syphilis Communicated by Leeches', *Lancet*, X/240 (1828), p. 14.

7 BIOMEDICAL LEECH

1 Roy T. Sawyer, 'Editorial', *Blood Coagulation & Fibrinolysis*, II (1991), pp. 63–4.
2 A. E. Shipley, 'Leeches', *British Medical Journal* (1914), pp. 916–19.
3 'When Leeches were Used for Black Eyes', *Times* (17 April 1965), pp. 10.
4 'Use of Leeches on Stalin Amazes U.S. Doctors', *Victoria Advocate* (5 March 1953), p. 2.
5 'Leeches and Penicillin', *Milwaukee Journal* (14 March 1953), p. 7.

6 Marvin E. Aronson, 'Leechcraft', *Journal of the Royal Society of Medicine*, xciv (2001), p. 372.

7 R. D. Montgomery, 'Leechcraft', *Journal of the Royal Society of Medicine*, xciv (2001), p. 553.

8 Charles Forte, *Forte: The Autobiography* (London, 1987), p. 46; 'Doctors Tried to Cure Andrea Bocelli's Blindness with Leeches', *Sunday Telegraph* (31 October 2010), p. 8.

9 'The Animals of Medicine', *British Medical Journal* (1911), p. 708.

10 Jim Endersby, *A Guinea Pig's History of Biology* (London, 2007).

11 E. M. Tansey, 'Henry Dale and the Discovery of Acetylcholine', *Comptes Rendus Biologies*, cccxxix (2006), pp. 419–25.

12 Lucy D. Leake, 'The Leech as a Scientific Tool', *Endeavour*, 2 (1983), pp. 88–93.

13 J. B. Haycraft, 'Über die Einwirkung eines Sekretes des offiziellen Blutegels auf die Gerinnbarkeit des Blutes', *Arch Exp Pathol Pharmakol*, 18 (1884), pp. 209–17.

14 'Hirudin', *Journal of the Chemical Society*, lxxxviii (1905), p. 339.

15 John Abel, Leonard Rowntree and B. Turner, 'On The Removal of Diffusible Substances from the Circulating Blood of Living Animals by Dialysis', *Journal of Pharmacology and Experimental Therapeutics*, v (1914), pp. 275–316.

16 J. Termier, 'Grenoble-Abortive Treatment of Surgical Phlebitis with Early Leaving of Bed', *Procès-verbaux et Mémoires*, 31 (1922), p. 949.

17 Howard Mahorner and Alton Ochsner, 'The Use of Leeches in the Treatment of Phlebitis and the Prevention of Pulmonary Embolism', *Annals of Surgery*, xcviii (1933), pp. 408–21.

18 Götz Nowak and Karsten Schör, 'Hirudin: The Long and Stony Way from an Anticoagulant to a Recombinant Drug and Beyond', *Thrombosis and Haemostasis*, xcviii (2007), pp. 116–19.

19 Bengt I. Eriksson et al., 'A Comparison of Recombinant Hirudin with a Low-Molecular-Weight Heparin to Prevent Thromboembolic Complications after Total Hip Replacement', *New England Medical Journal*, cccxxxvii (1997), pp. 1,329–35.

20 Roy T. Sawyer, 'A Sanguine Attachment: 2,000 Years of Leeches in Medicine', in *Medical and Health Annual, 1999* (London, 1998), pp. 88–103.
21 Jack McClintock, 'Bloodsuckers', *Discover Magazine* (December 2001).
22 Roy T. Sawyer, 'Johann Friedrich Dieffenbach: Successful Use of Leeches in Plastic Surgery in the 1820s', *British Journal of Plastic Surgery*, LIII (2000), pp. 245–7.
23 Johannes Esser, *Artery Flaps* [1928] (Rotterdam, 2003).
24 M. Derganc and F. Zdravic, 'Venous Congestion of Flaps Treated by Application of Leeches', *British Journal of Plastic Surgery*, XIII (1960), pp. 187–92.
25 J. Baudet, 'The Use of Leeches in Distal Digital Replantation', *Blood Coagulation and Fibrinolysis*, 2 (1991), pp. 193–6.
26 Kenneth Gosling, 'Modern Medicine – With Leeches', *The Times* (17 March 1982), p. 1.
27 'Doctors Combine Modern Surgery, Ancient Leeching to Save Boy's Ear', *Evening Herald (Rock Hill)* (25 September 1985), p. 11.
28 Emily E. Lazarou et al., 'The Psychological Effects of Leech Therapy After Penile Auto-amputation', *Journal of Psychiatric Practice*, XII (2006), pp. 119–23.
29 Alan Park, 'The Case of the Disappearing Leech', *British Journal of Plastic Surgery*, XLVI (1993), p. 543.
30 A. MacQuillan, 'Taking a Leech to Blood: But Can You Make Him Drink?', *British Journal of Plastic Surgery*, LV (2002), pp. 540–41.
31 Anders Baerheim and Hogne Sandvik, 'Effect of Ale, Garlic, and Soured Cream on the Appetite of Leeches', *British Medical Journal* (1994), p. 1,689.

CONCLUDING LEECH

1 Emilie Autumn, *A Bit O' This & That* (Trisol, 2008).
2 Kira O'Reilly, *Bad Humours / Affected* (1998), recording available at the Live Art Development Agency, Rochelle School, London.

3 Kira O'Reilly, 'One Hundred Wound Sites or More', *a-n Magazine* (April 2001), p. 26.
4 Theodor Adorno, *Minima Moralia: Reflections from Damaged Life* [1951], trans. E.F.N. Jephcott (London, 1974), p. 105.
5 Michel Serres, *The Parasite,* trans. Lawrence R. Schehr (Minneapolis, MN, 2007), p. 8.
6 Ibid., p. 13.

Select Bibliography

Anderson, Katherine, *Predicting the Weather: Victorians and the Science of Meteorology* (Chicago, IL, 2005)

Braunstein, Jean-François, *Broussais et le matérialisme: médecine et philosophie au XIXe siècle* (Paris, 1986)

Broussais, F.-J.-V., *Leçons du docteur Broussais, sur les phlegmasies gastriques, dites fièvres continues essentielles des auteurs, et sur les phlegmasies cutanées aiguës* (Paris, 1819)

Chandra, Mahesh, *The Leeches of India: A Handbook* (Calcutta, 1991)

Elliot, J. M., and K. H. Mann, *A Key to the British Freshwater Leeches, with Notes on their Life Cycles and Ecology* (Ambleside, Cumbria, 1964)

Horn, George, *An Entire New Treatise on Leeches, Wherein the Nature, Properties, and Use of That Most Singular and Valuable Reptile, is Most Clearly Set Forth* (London, 1798)

Fermond, Charles, *Monographie des sangsues médicinales: Contenant la description, l'éducation, la conservation, la reproduction, les maladies, l'emploi, le dégorgement et le commerce de ces annélides suivie de l'hygiéne des marais à sangsues* (Paris, 1854)

Gruner, Oskar Cameron, *The Canon of Medicine of Avicenna* (London, 1930)

Hoeppli, R., *Parasites and Parasitic Infections in Early Medicine and Science* (Singapore, 1959)

Johnson, James Rawlins, *A Treatise on the Medicinal Leech: Including its Medical and Natural History, with a Description of its Anatomical Structure: also, Remarks Upon the Diseases, Preservation and Management of Leeches* (London, 1816)

Lemm, Vanessa, *Nietzsche's Animal Philosophy: Culture, Politics, and the Animality of the Human Being* (New York, 2009)

Mann, Kenneth H., *Leeches (Hirudinea): Their Structure, Physiology, Ecology and Embryology* (Oxford, 1962)

Merryweather, George, *An Essay Explanatory of the Tempest Prognosticator in the Building of the Great Exhibition for the Works of Industry of All Nations* (London, 1851)

Michalsen, Andreas, Manfred Roth and Gustav Dobos, *Medicinal Leech Therapy* (Stuttgart/New York, 2007)

Muller L. G., *Der Medizinische Blutegel ('Hirudo medicinalis') Oder Naturhistorische Beschreibung des Blutegels* (Quedlinburg, 1830)

Nachtrieb, Henry Francis, John Percy Moore and Ernest E. Hemingway, *The Leeches of Minnesota* (Minneapolis, MN, 1912)

Negus, Robert P., *Essay on Leeches: A Practical Hand Book* (Melbourne, 1868)

Price, Rees, *A Treatise on the Utility of Sangui-Suction, or Leech Bleeding in the Treatment of a Great Variety of Disease* (London, 1822)

Sawyer, Roy T., 'Why we need to save the medicinal leech', *Oryx*, XVI (1981)

—, *Leech Biology and Behaviour, Volume One: Anatomy, Physiology and Behaviour* (Oxford, 1986)

—, *Leech Biology and Behaviour, Volume Two: Feeding, Biology, Ecology and Systematics* (Oxford, 1986)

—, *Leech Biology and Behaviour, Volume Three: Bibliography, Biology, Ecology and Systematics* (Oxford, 1986)

—, 'A Sanguine Attachment: 2,000 Years of Leeches in Medicine', *Medical and Health Annual 1999* (London, 1998), pp. 88–103

—, 'The Trade in Medicinal Leeches in the Southern Indian Ocean in the Nineteenth Century', *Medical History*, XLIII (1999), pp. 241–5

Serres, Michel, *The Parasite*, trans. Lawrence R. Schehr (Minneapolis, MN, 2007)

Vayson, Louis, *Guide pratique des éleveurs de sangsues* (Paris, 1855)

READING FOR CHILDREN

Blaxland, Beth, *Annelids: Earthworms, Leeches, and Sea Worms* (Philadelphia, PA, 2002)

Epstein, Sam, and Beryl Williams Epstein, *You Call That a Farm? Raising Otters, Leeches, and Other Unusual Things* (New York, 1991)

Halton, Cheryl H., *Those Amazing Leeches* (Upper Saddle River, NJ, 1990)

Kite, L. Patricia, *Leeches*, Early Bird Nature series (Minneapolis, MN, 2004)

Neuman, Pearl, and John W. Reynolds, *Bloodsucking Leeches* (New York, 2009)

Schaefer, Lola, *Leeches*, Ooey-Gooey Animals series (Oxford, 2003)

Silverstein, Alvin, Virginia Silverstein and Laura Silverstein Nunn, *Dung Beetles, Slugs, Leeches, and More: The Yucky Animal Book*, Yucky Science series (Berkeley Heights, NJ, 2010)

Somervill, Barbara A., *Leeches: Waiting in the Water* (New York, 2008)

Associations and Websites

BIOPHARM LEECHES, THE BITING EDGE OF SCIENCE

Established in 1984, Biopharm is an international company based in Hendy, South Wales, producing medicinal leeches for use in plastic and reconstructive surgery worldwide.
www.biopharm-leeches.com

INTERNATIONAL MEDICAL LEECH CENTRE

Established in 1937, the International Medical Leech Centre, based in Moscow, Russia, offers training programmes in hirudotherapy, conducts research into the therapeutic action of *Hirudo medicinalis* and produces leeches for medicinal use.
www.leech.ru/en

RICARIMPEX SANGSUES MÉDICINALES

Established in 1835 by Béchade, this Bordeaux-based company has been in continuous trade up to the present day. RICARIMPEX continues to supply *H. medicinalis* for medical use across the world.
www.sangsue-medicinale.com

THE LEECH LAB (LABORATORIES OF PHYLOHIRUDINOLOGY)

Research group led by Mark E. Siddall, based at the American Museum of Natural History, New York, investigating the taxonomy, systematics, phylogeny and evolution of leeches.
www.research.amnh.org/~siddall

Acknowledgements

We would like to thank our colleagues at CHSTM (University of Manchester), many of whom have listened politely to curious leech stories, even when they were told over lunch. We appreciate their fortitude when, in order to truly know leeches, we invited ten retired medicinal leeches to join us at CHSTM. In particular, we owe a debt to Mick Worboys, who not only tolerated our desire to develop a stray sentence into a book-length project, but actively encouraged the pursuit of an interesting and unusual history. We would also like to thank Matthew Cobbe who, in addition to feeding us leech information whenever he encountered it, was also brave enough to allow us to introduce our leechy friends to the interested public. Numerous audiences have listened to, responded, and encouraged our interest in leeches – to all we are grateful. But few more so than Roger Cooter, whose forceful insistence that the world needed to know how leeches of the past have aided problematic prostate glands did much to convince us of the need for this book. We would also like to thank the audience at the 'Future of the History of Medicine Conference' (UCL, London, 2010), who believed us when we said the future of history was to look beyond the human. In the course of writing we accumulated too many debts to too many people, but would like to acknowledge the encouragement of Kathryn Ashill, Emilie Autumn, Donald Blaufox, Mike Brown, Kat Foxhall, Teunis W. van Heiningen, Emma Jones, Melissa King, Andrew Knight, Bill Luckin, Kira O'Reilly, Rick Veitch, Kris Weller and Duncan Wilson. At Reaktion, we are grateful to Jonathan Burt, Michael Leaman and the editorial staff for all their assistance.

As ever, we are indebted to the Wellcome Trust, whose generous support (through grant 084988/z/08/z) contributed to the realization of this work.

The last word, however, must go to leeches everywhere, for it is with them that our greatest gratitude lies. Without leeches, our lives over the past years would have been dramatically impoverished. Wherever you are, may you be loved.

Photo Acknowledgements

Used with courtesy of the artist, Kathryn Ashill: p. 181; used with kind permission of Emilie Autumn: pp. 71, 171; © Capcom, used with permission: p. 136: Andrew Knight (www.andrewadventures.info): p. 98; Manuel Krueger-Krusche: p. 25; Mary Evans Picture Library: p. 85; Reproduced with the permission of M. Donald Blaufox MD, PhD, from the Museum of Historic Medical Artifacts (http://www.mohma.org): p. 69, 112 left; National Geographic Stock: pp. 169 (Ira Block), 6 (Jason Edwards), 29 top (Bianca Lavies); National Institutes of Health, Bethesda, Maryland: p. 11; Kira O'Reilly, *Bad Humours/Affected* (1998), Bonnington Gallery, Nottingham Trent University (Photography, John Morgan), used with permission of the artist: p. 180; Rex Features: p. 95 (Roger-Viollet); Shutterstock: pp. 75 (Antonio Abrignani), 67 (Tatiana Belova), 47 bottom, 153 (Mircea Bezergheanu), 14 (Sergey Goruppa), 28, 117 (James Harrison), 35 (David Lade), 107 (Patrick Landmann), 122 (Hugh Lansdown), 89 (Semen Lixodeev), 37 (Sergey Lukyanov), 149 (Galina Mikhalishina), 31 centre, 36 bottom, 48 (Morphart Creations Inc.), 44 (mrfiza), 152 (Natursports), 154 (Photocrea), 60 (Real Illusion), 128 (riekephotos), 36 top (Carolina K. Smith, MD), 121 (sydeen), 10, 119 (szefei), 43 (Rudy Umans); Science Photo Library: pp. 38 (Eye of Science), 125 (Scientifica, Visuals Unlimited), 30, 165 (Geoff Tompkinson); Science and Society Picture Library: pp. 92 (Richard Bosomworth), 164 (Daily Herald Archive); Swindon Collection, Swindon Libraries: p. 82; © The Wellcome Trust, London: pp. 17, 18, 41, 53, 55, 56, 57, 62, 68, 74, 76, 78, 79, 80, 81, 86, 90, 102, 113, 120, 156, 159, 33 (Mark de Fraeye), 158 (Richard Wingate); By permission of Whitby Literary and Philosophical Society: p. 105; © 1989 Rick Veitch: p. 133.

Index